POWER TO THE CITY

THE HISTORY OF

THE
EXETER ELECTRIC LIGHT & POWER STATION

Haven Road, Exeter

A building saved from demolition

Dick Passmore

First published in Great Britain by Little Silver Publications

Copyright © Dick Passmore 2008

The right of Dick Passmore to be identified as the author of this work is hereby asserted.

ISBN 0–9544472–5–5

Little Silver Publications, Little Silver, Matford, Exeter EX2 8XZ

CONTENTS

INTRODUCTION

Adjacent to the River Exe, alongside Exeter's Canal Basin, stands an interesting brick and stone building which, because it has been virtually unused for many years, is not readily noticed by tourists or passing residents. Much of the original structure has long since gone, but nevertheless that which remains is an attractive building. After years of neglect, it will soon be enjoying a new lease of life thanks to a well thought out and sympathetic restoration project by Millhouse Partnership, and is a building worthy of having its history set in print for future generations.

Many people who have never before had the opportunity to venture inside this building will be amazed at the unusual and fascinating interior – fascinating, even though it was originally constructed merely as an electricity power station.

What remains is, importantly, the façade and a major section of the interior. The building has an important part to play in Exeter's history, for it was from here that electricity was generated to supply the city from 1905 until 1960, and also to supply power to the Exeter Corporation Tramways for almost thirty years. The days of the trams are now distant, but the building that provided their electricity is just beginning resurgence. The days of electricity supplied by steam-driven generators also seems a long time ago, for today more modern methods, including nuclear power, have taken on this rôle.

In this book, the reader is given a brief history of the Exeter Electric Light & Power Station, and why the building is in the position it occupies. Furthermore, without being a scientific journal by any stretch of the imagination, the book tells how electricity has evolved, how it came to Exeter, and the subsequent incredible increase in demand.

The author was fortunate to have been given access to numerous photographs taken by amateur photographers at the beginning of the 20th century, during the early days of photography in this country. Some of these photographs have never been published before. In several cases, due to their age, they are of poor quality, and have not necessarily reproduced well. However, it is more important that they have been retained in a suitably professional archive, created by electrical historians, that will ensure their posterity.

THE ADVENT OF ELECTRICITY

The invention of electricity cannot be attributed to any one person, simply because it was never "invented". The word electricity merely relates to various concepts of energy, and by definition it is *any phenomenon associated with stationary or moving electrons, ions, or other charged particles*. Even the Ancient Greeks and Babylonians knew of its presence, albeit in various forms and not then called electricity, although their knowledge was not as specific and detailed as that of today's scientists.

Over the centuries, many people have been intrigued by this phenomenon and, by understanding and making use of it, have adapted it for our day-to-day purposes. A flash of lightning is natural electrical energy in a form that we see lighting up the sky; an electric light bulb, quickly turned on and off, is electrical energy that momentarily lights up the surrounding area, and is therefore exactly the same – but man-made because of our increasing knowledge of the original phenomenon.

We take electricity for granted in the twenty-first century, and it does, of course, affect our everyday lives – something that, today, we could not live without, although until comparatively recently we did! It is a fact that there are still a few domestic properties in some parts of the British Isles that exist without electricity, and many more that have no mains electricity, but rely solely on a generator or some means of providing electrical power. Indeed, in August 2008, a property in Wales was connected to electricity for the very first time, having previously been lit by candles and gas. For centuries, the only indoor lighting was by way of rush lights, tallow-dips, wax candles, oil lamps or various other means of creating light using naked flames. Whilst heat, light, and power are probably the three main uses to which we put electricity in our homes and businesses, there is a lot more to electricity than merely throwing a switch or turning a control. To understand it in totality requires a scientific knowledge that is probably far beyond most of us.

Such in-depth knowledge need not be investigated here, although it is necessary to delve a little into how electricity has evolved, for without it, of course, there would have been no need to build power stations.

Various discoveries in the world of electricity have, over the centuries, been connected to many well-known names. Michael Faraday is known for his work on electro-magnetic induction and the development of magnetic fields. In 1752, Benjamin Franklin experimented in a thunderstorm, when he flew a kite with a key attached to the string, resulting in a spark that proved lightning was an electrical charge. A brave man! The surnames of André-Marie Ampère, Georg Ohm and Samuel Morse have all left their individual mark in the world of electricity, whilst others, such as Graham Alexander Bell, Joseph Swan and Thomas Edison, have also been well documented as a result of their experimental electrical work. Some two hundred years ago, Humphrey Davy passed an electrical current across two platinum strips, and produced a glow, although it only lasted for a few seconds as a light source, due to the strips evaporating too quickly.

Of all these names, it is perhaps Michael Faraday, Thomas Edison and Joseph Swan who are the most important in relation to this particular story.

Michael Faraday was born in Surrey in 1791. As a youth he was an apprentice book-binder in London, but whilst still in his teens he developed a fascination for science, attending lectures and keeping notes of anything he considered of scientific interest. When aged twenty-one, Faraday attended four lectures by Sir Humphrey Davy at the Royal Institution, and his life changed after listening to them.

He eventually left his chosen trade and took a position as laboratory assistant at the Royal Institution. He worked in close association with Sir Humphrey Davy, and was invited to join Davy on his travels around Europe in 1813, taking notes and generally creating manuscripts of Davy's work. On returning to the Royal Institution, Faraday started to experiment in various aspects of science, in particular electro-magnetism. When Faraday discovered this phenomenon he went on to construct the first, albeit rather crude, dynamo, which he demonstrated in 1831. The rest really is history, for Michael Faraday was to become one of this country's most famous scientists and inventors. Faraday's importance to science was such that Queen Victoria expressed her gratitude by making him a gift of a splendid mansion near Hampton Court, and it was in this house that Faraday died in 1867.

Thomas Alva Edison was born in Ohio, USA, in 1847, the seventh child of Canadian Samuel Edison and his wife Nancy – although the family was of Dutch, not Canadian, origins. Unbelievably, Edison had very little formal education, being taught largely by his mother. An inquisitive but bright youth, his quest for knowledge on just about any subject was never satisfied. Even when he lost his hearing in one ear, and became profoundly deaf in the other, he carried on, determined to succeed in some field, although he really had no real idea of what that would eventually be. It is believed that Edison's hearing problems were the result of suffering from scarlet fever as a child, coupled with frequent ear infections afterwards.

He created a newspaper when a mere youth, being allowed to sell it on a local railway where he was employed, and in his teens he sold just about anything he could in addition to his newspaper, including snacks, and fruit and vegetables.

At the age of sixteen he started work in a telegraph office, and his fascination for science increased even more, even though he was virtually devoid of the usual standard of education other children of his age had received.

Edison became interested in Faraday's scientific projects and research work, and this inspired him to involve himself in similar experiments. The result was that he became an incredible inventor with well over one thousand patents credited to his name. From his research laboratory in New Jersey, Edison was able to make major contributions to telegraphy, the development of telephones, and, of course, the incandescent lamp. Yet it is not possible to be sure just how many inventions can be attributed to Edison himself. It is known that Edison employed many assistants, and several of these were scientists of considerable ability. It is possible – and has been claimed previously by historians – that Edison could well have made claim to certain inventions that were the work of his assistants.

Joseph Wilson Swan was born in Sunderland, in 1828. In his early career he was apprenticed to John Mawson, a pharmacist (and his future brother-in-law), and later became a partner in a firm of manufacturing chemists, *Mawson's*, in Newcastle upon Tyne. Although the premises are still in existence, the company ceased trading in the early 1970s.

As well as being a chemist, Swan also had a general interest in other sciences. For many years he experimented in various aspects bromide printing paper (familiar to photographers), cellular lead plate storage batteries, artificial cellulose thread (the prototype of artificial silk) and the carbon process, but probably spent most of his time experimenting with the electric filament lamp – as did several other scientists in this country and in America.

In 1878, Joseph Swan invented an incandescent carbon filament lamp (which can been seen in London's Science Museum) and he demonstrated this at a meeting of the Newcastle upon Tyne Chemical Society on the 18th December of that year. Two years later, Sir William Armstrong had a small hydro-electric plant constructed at his home, *Cragside*, in Northumberland, and engaged Swan to light the picture gallery. This was almost certainly the first dwelling house in the world to be lit by incandescent electric lamps.

By 1881, Joseph Swan had set up the Swan Electric Lighting Company.

Also in 1878, Edison had been working on his version of the incandescent lamp using platinum, but this was not to Edison's satisfaction, because platinum had insufficient resistance for Edison's requirements, and so he began to experiment with carbon. This was to prove far more successful, and in October 1879 Edison was wise enough to take out a patent on this invention. In February of the following year the first Edison commercial incandescent lamps were demonstrated in London. In October the first public demonstration of electric lighting on a large scale by means of incandescent lamps was given in Newcastle.

At about the same time as Swan set up his company, Edison had created the Edison Electric Light Company in Pearl Street, New York, and opened a generating station there that was capable of supplying 110 volts of direct current to a few households in Manhattan. In the same year, he opened a steam-generating power station in Holborn, London, that was capable of supplying electricity to street lamps and houses in the immediate area.

In addition to Edison's contributions in helping others in the development of the phonograph, telegraphy and the telephone, by 1900 he had extended his own experimenting to electrical generating equipment, storage batteries and even motion pictures.

Unfortunately, Swan had had decided not to patent his invention *(see later note)* and when Edison came up with his similar device, which did have a patent, Swan was far from happy, and there developed an incredible argument between the two scientists. It eventually culminated in the two men agreeing to a deal, and together they formed a business known as the Edison & Swan United Light Company, located in a factory in Ponders End, near Enfield in Middlesex. Because they had developed improved vacuum technology, Edison and Swan were able to use carbon where other researchers had failed. The lamps they produced were known as

Royal Ediswan Lamps, due to the pair having obtained a Royal Warrant.

The experiments carried out by another scientist, Humphrey Davy, resulted in Davy's invention of the first basic arc lamp in 1808, although others went to on design much more sophisticated and successful arc lamps by the 1870s. The carbon rods only lasted for around eight hours, and so the more popular lamps had two pairs of carbon rods, and as the first pair burned away, so the second pair came into operation – doubling the length of time the lamp was illuminated. Whilst this type of light was ideal for large outdoor areas, particularly in lighting streets, it was not suitable for domestic use, as the lamps were far too bright, and they also gave off a rather unpleasant smell.

Humphrey Davy was born in Penzance, Cornwall, in 1778. He attended schools in his home town, and also in Truro, before taking an apprenticeship with a surgeon and apothecary in Penzance. He remained in Cornwall until the age of twenty, but even though he moved away he never lost touch with his home, or his school, and indeed on his death he left a considerable endowment to benefit the pupils of his alma mater. By the age of twenty-three, Davy had been made Professor of Chemistry at the Royal Institution, and in 1820 he became president of the Royal Society, and was re-elected to that position for seven consecutive years, until failing health forced him to step down. He died in Geneva at the age of just fifty. Despite his work in various fields of science, particularly his work in electrical conductivity, Davy is, perhaps, best known for the invention of the miner's safety lamp. Davy was created a Baronet in 1818.

Davy's early experiments in lighting had been unsuccessful because the platinum strips he used had not been encased, thus allowing the surrounding air to destroy them. The incandescent bulbs of both Swan and Edison were similar to that in Davy's experiments, but they used platinum filaments that were encased in a vacuum. The vacuum was made by means of a sealed glass bulb and this prevented the platinum absorbing hydrogen from the atmosphere, which caused it to develop cracks and weaken. When an electric current was passed across the filaments, a light source was obtained, and was not destroyed by any hydrogen. This was the forerunner of today's electric light bulb, now an everyday item taken for granted throughout the world.

There can be little doubt, however, that one of the most useful inventions of both Joseph Swan and Thomas Edison – certainly as far as the general public worldwide is concerned – was the invention of the incandescent light bulb, for without this aspect of their contribution to the world of electricity, our own world would, perhaps, seem a most strange – and indeed a much darker – place.

Note: Davy, Faraday and Swan never patented their inventions or discoveries, as they considered that they were "carrying out their work for the good of mankind". Edison, on the other hand, was more commercially minded.

The bulb used by Edison in experiments at Menlo Park, California, in December 1879.

Image released under GNFL, with author's permission.

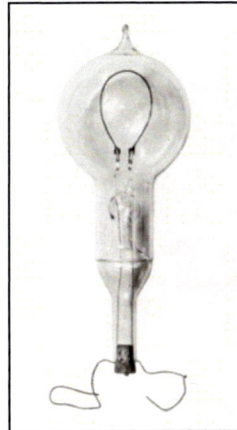

Edison's carbon horseshoe bulb.
This shows the platinum filaments encased in glass to form a vacuum.

Photograph courtesy Thomas A. Edison Papers @ Rutgers, State University of New Jersey.

In August 2008, *The Daily Telegraph* reported that "a Swan Edison filament has blazed for more than seventy years at a house in Cowes, Isle of Wight." The bulb had been purchased in 1938. It also states "the average standard *[filament]* bulb will burn for between 750 and 1,000 hours, although it is estimated that this bulb had possibly carried on for in the region of 600,000 hours!"

Thomas A. Edison.
1847 – 1931

Photograph courtesy of
Thomas A. Edison Papers
@ Rutgers, State University of New Jersey.

Joseph Wilson Swan.
1828 – 1931

Photograph courtesy of
Discovery Museum,
Tyne and Wear Museums.

ELECTRICITY COMES TO EXETER

It was during the latter part of the nineteenth century that various people, including Edison, were finding that their experiments with electrical energy were proving successful. By the 1870s, generators were being manufactured, and these were designed to create a continuous supply of electricity. Arc lamps had been developed to work from this supply, resulting in the provision of considerable light, virtually wherever it was required. Arc lamps are so named because the illumination they provide is the result of an electrical charge "arcing" across two conductors placed close to each other.

Because of the amount of light these lamps could provide, they became very popular for use in large buildings such as factories and warehouses, although for domestic use they were generally considered too bright. Soon, public buildings all over the country were being lit by means or arc lamps, and one of the main suppliers was the Crompton Company – said to be the most popular supplier of arc lamps in the 1880s, even though the company had only been founded in 1878. Crompton was so successful that he was engaged to supply lighting to Windsor Castle and Holyroodhouse Palace.

At the end of the nineteenth century, electricity came to Exeter for the first time in the form of a demonstration of this new phenomenon by the Devon & Cornwall Electric Light and Power Company, during the "Olde English Fayre", an exhibition at the Victoria Hall, Queen Street in July 1882. For various reasons, this company was short-lived, and went into liquidation in January 1883. However, following the demise of that company, a similar business, The Union Electric Light Company, attempted to secure the rights to supply electricity to Exeter, but these attempts were rejected by Exeter City Council.

The Victoria Hall, Exeter, after the disastrous fire in 1919.

The Victoria Hall was the setting for the "Old English Fayre" in 1882.

Nevertheless, six years after the Victoria Hall exhibition, it was a shoemaker by the name of Henry Massingham who persuaded Exeter City Council to allow him to stage a much bigger demonstration of what electricity could do!

Henry George Massingham was a boot and shoe manufacturer who originated from Bath, although his company had various outlets throughout the southwest of England. He was not an engineer, and he knew comparatively little about electricity, but he had seen the possibilities with his own eyes when Bristol Cathedral was illuminated in the late 1800s, and he enthused about the future for this means of illumination.

Fascinated by what he had seen in Bristol, he was astute enough to realise the potential of electricity in towns and cities, his enthusiasm being shared by others who supported him. Following successful demonstrations in Taunton, he successfully tendered for the supply of electric lighting in a large part of Somerset's capital town. This gave him the opportunity of joining forces with other businessmen in the town, and, in 1885, together they formed the Taunton Electric Lighting Company Ltd., of which Henry Massingham was Managing Director. This was the first time a public supply of electric light had been made available in the southwest of England.

Such was the success of this company that enquiries came from far and wide, some towns and cities even sending their representatives to view the set-up at first hand. Whilst many were still rather sceptical and unsure about electricity, Massingham forged ahead, travelling further afield to show businessmen in Exeter, Bristol, Bath and Weston-super-Mare and other locations exactly how electricity could provide both light and power for their companies and their homes. Fortunately, the authorities in Exeter were able to see possibilities in Massingham's demonstrations, but they also needed to be convinced that this was something into which they should invest, and that was not easy. Whilst the authorities realised that it could be an important step forward for the City's future, they were also well aware that any decision to go ahead could also be a disaster.

Massingham was aware that the council was not yet wholly convinced that they should go ahead with any lighting scheme, and in December 1887, he persuaded them to allow him to carry out further demonstrations in Exeter, and these took place during February of 1888, lasting the whole month. According to a report in *The Western Morning News*, they were "a decided success". The experiment included setting up arc lamps on high poles at various locations in the centre of the city – exactly as he had done elsewhere – to show how electricity could be conveyed from the generating plants to houses and businesses at any given location. The electricity was provided as a direct current, using American Thomson-Houston dynamos to supply the arc lamps used in the demonstration.

Some of the businesses that enjoyed being supplied with electric light during this experiment included The New London Hotel in London Inn Square; The Arcade Coffee Tavern, and Trowells the jewellers, in the Eastgate Arcade; J.H. Newman & Co., and The Half Moon Hotel – both in High Street; The Royal Albert Memorial Museum in Queen Street; The Spirit Vaults in Sidwell Street; The Athenaeum in Bedford Circus, and several other businesses in the centre of the city.

Massingham was so confident that his ideas would be accepted, that he negotiated the purchase of a site at Trew's Weir, in the St Leonard's area of the city, and adjacent to the River Exe. This was where his Company intended to install four 24-horse power turbines using some 3,000 cubic feet of water per minute to power the turbines, rather than a steam system.

Whilst the city council was still unable to make up its mind, there can be no doubt that Massingham had achieved a degree of success as a result of the arc lamp experiments in the city. Following their observations of his experiments, those businessmen with foresight must have been satisfied that here was something that they could not afford to ignore, for shortly afterwards several of them decided to join Massingham in creating the Exeter Electric Light Company.

However, it was not as straightforward as the directors would have wished. The Exeter Electric Light Company was incorporated following Massingham's experiments, and during March of 1888, two thousand shares were offered at £10 per share, with the intention of realising £20,000. Unfortunately, there was not sufficient public interest, and the few applications that had been received were returned, with the shares being allotted to the directors themselves.

It was with considerable regret to Massingham that his new company had started its existence by being under-capitalised. Despite the efforts of directors, public interest continued to be poor, and in the next three years only one thousand two hundred shares were taken up. As a result, the company found it necessary to issue £10,000 of 5% debentures to raise the necessary money for it to expand.

The Eastgate Arcade, High Street.

Located in Exeter's High Street, and destroyed during the Second World War.

The New London Hotel, London Inn Square.

Northernhay Place is seen on the left, and the former Theatre Royal in the background on the right. The hotel was owned by Robert Pople.

Meanwhile, within the city council some of the more influential councillors were backing proposals to continue lighting the streets by means of gas, but others (who were equally influential) wanted the new system of electricity. It is perhaps of interest at this point to note that the then Mayor, Charles Roberts, was a solicitor who, in the course of his profession, acted for The Exeter Electric Light Company, whilst Alderman Ellis was Chairman of that company. At the same time, Alderman Richard Daw was also a solicitor, and he acted for the Gas Company!

With these apparent conflicts of interest, it is quite easy to understand why a "power struggle" developed in connection with Massingham's proposals which had been forwarded to the council, although it would appear that in those days, unlike modern-day procedure, there seems to have been no obligation to "declare an interest" in the subject being debated, whereby the interested parties take no part in proceedings during the debate.

Seven tenders for the supply of electric lighting within a central area of the City were submitted to the Council, and that of the Exeter Electric Light Company was the lowest. Even so, there was still cause for concern as that tender amounted to almost £1,500 per annum to cover the same area that could be supplied by gas for just under £700 per annum – a distinct problem for the 'electric lobbyists' within the Council. There was, however, one major difference between the two systems, and that was the amount of actual illumination available. Massingham was able to legitimately show that even though the electric system was more expensive, and *pro-rata* twice the price of gas, the overall illumination offered by his electrical system was over fifteen times greater than that of the gas lighting – a huge difference.

Yet even that difference was not enough to sway the Council, and due to the vast discrepancy in the overall costs, not one of the tenders was accepted.

Massingham's company was not totally beaten however, and the directors resolved to continue, despite the knowledge that they would have no guarantee of any income from a contract with Exeter City Council. Added to the rejection of their tender, the Company's plan to use the Trew's Weir site was also rejected by the Council, on the advice of the Borough Surveyor and the City's Wharfinger (the manager of that area of the River Exe where boats were loaded and off-loaded), both of whom were key figures of the Council, and also members of the Navigation Committee of the Council, so thus having considerable sway.

This further rejection compelled the Company to find a suitable alternative location if they were to set up their company. In February 1889, they were able to purchase the freehold of premises known as the Rockfield Works in New North Road, a short distance from the very centre of Exeter, and a building that they considered eminently suitable for their purposes.

Whilst the city council was still unable to make up its mind, there can be no doubt that Massingham had achieved a degree of success as a result of the arc lamp experiments in the city. Following their observations of his experiments, those businessmen with foresight must have been satisfied that here was something which they could not afford to ignore, for shortly afterwards several of them decided to join Massingham in creating the Exeter Electric Light Company.

Twelve months of frustration and rejection had not deterred the company's directors, and once the purchase of Rockfield Works was complete, they decided to go ahead with the installation of the machinery necessary for them to be able to supply electricity from their new premises.

Without the backing of the Exeter City Council, a body that was destined to become the major user of electricity in the city, Massingham and his fellow directors had taken a huge step forward, even if to many observers that step appeared to be slightly foolhardy.

An early photograph of Henry Massingham.

Photo courtesy SWEHS.

THE ROCKFIELD WORKS

The generating equipment needed for this new operation was, as mentioned earlier, housed in a building known as the Rockfield Works, in New North Road, adjacent to the main railway line at Exeter Central Station. The building remained in being until the end of the twentieth century, when it was demolished to make way for the modern Longbrook House complex.

The premises in New North Road were in the ownership of Mr William Edwards, where he operated his business as a Silk Hat and Cap Works. During the latter part of 1888, the premises were advertised for sale. The estate agent's description of the building included the fact that it was *most suitable, or capable of being adapted, for any large manufacturing business.*

The Rockfield Works was said to have a frontage of just over one hundred feet (thirty metres) and a depth of ninety feet (twenty-seven metres), and it contained a workable area of something approaching eight thousand square feet. There was also a basement area about half of that size. As far as Henry Massingham was concerned, it certainly seemed a most appropriate building for his purposes, being adjacent to a railway station, and close to the city centre. The railway was needed to bring in coal supplies, and also water to create the steam to drive the generators. Being close to the city centre was a bonus, as that made it easier to lay his cables supplying electricity.

When the Exeter Electric Light Company started operations at the New North Road premises, the equipment installed to supply the city with electricity comprised the following:

- ❑ Two Babcock & Wilcox water tube boilers, each capable of 150 horsepower capacity.

- ❑ One Worthington feed pump, and two injectors.

- ❑ Two Fowler horizontal engines, each capable of 150 horsepower capacity.

- ❑ Two Westinghouse 25hp triple expansion engines.

- ❑ Two Brush-Morley 75Kw alternators – rope driven, for AC (Alternating Current) supplies.

- ❑ Two Thomson-Houston dynamos – for DC (Direct Current) supplies.

In addition, there were two alternating current generators, plus two with direct current, capable of providing a distribution system by underground and overhead cables that would eventually supply most of the city with electricity. The total cost of this equipment, and all other associated costs, was in the region of £21,000, a considerable sum of money at that time.

The actual supply of electricity from the Rockfield Works commenced in November 1889, but before that could happen the distribution system needed to be set up. At the start, poles were permitted to carry the supply cables overhead, and one hundred cast-iron poles were erected from Exe Bridge, at the lower end of Exeter, to St Sidwells at the top end. This included poles in Bedford Street, Queen Street and New North Road These poles had a circular base, and were set some six feet (about two metres) into the ground, but protruding almost as much above ground. Into that base was dropped a thirty feet high pole, also cast iron, and the poles were spaced approximately one hundred yards apart. The supply cables were attached to the poles by means of a wrought-iron bracket, and generally there were either three or six cables, depending on the location and service required.

The Rockfield Works, New North Road.

Above, seen at the beginning of the twentieth century. The words "Exeter Electricity Company" can be seen at the top of the façade. Below, being used as a taxi office, health centre and garage business. towards the end of the same century.

Both photographs courtesy SWEHS.

The 1882 Electric Lighting Act contained clauses that stipulated cables should be *a minimum of twenty feet from the ground, and thirty feet where crossing a street, and at least six feet from any building (except when entering that building).* This requirement caused problems for the various Fire Brigades employed by Insurance Companies, when using ladders at fires in upper storeys, and also when using jump-sheets to catch people jumping from first-floor windows to escape a fire. It probably caused even more problems for the newly formed Exeter Fire Brigade, set up shortly as a result of the fire in Exeter's Theatre Royal in 1887, for they were obliged to attend fires at any building, not just those insured by any particular Insurance Company.

Due to these problems, the General Purposes Committee of the council asked the company to install isolating switches that would enable the Fire Brigade to isolate certain areas, in order that other areas away from any fire would not be cut off, and this was carried out.

However, shortly after the poles carrying electric cables around the city had been erected, there were more problems for the Electric Light Company as the poles were deemed unsightly, and caused a considerable number of complaints. Despite these numerous complaints and objections – including the council arguing that they actually had the power to order the removal of the posts – the company was still allowed to erect additional posts in South Street. The end came when the company applied for permission to erect even more posts for overhead cables, this time in Northernhay Place, where the Devon & Exeter Club required electric lighting in its billiard room. The council decided that granting permission could affect the public gardens in Northernhay, and permission was refused. Furthermore, the city council decided that all future cables should be installed underground, thus avoiding any additional poles, and eliminating further complaints. This was strongly contested by the company, largely on account of the cost involved, but it was unable to persuade the council to change its decision.

There was a requirement in the Electric Lighting Act of 1882, whereby any undertaking wishing to break open highways to lay cables underground would be required to obtain a Licence, or a Provisional Order, from the Board of Trade. A further stipulation was that the undertaking needed consent from the local authority in whose area they wished to lay underground cables. However, in the event of the local authority refusing to grant such permission, the undertaking could apply directly to the Board of Trade if it thought the refusal was not justified.

This actually happened in the case of the Exeter Electric Light Company. Following the council's decision, the company gave notice that it intended to apply for an Order on 25th May 1890. The council, meanwhile, managed to persuade the Board of Trade to promote the Exeter Electric Lighting Act, and this was given Royal Assent on 3rd July 1891. The passing of this Act gave the council considerable clout over the company, particularly by way of their power in being able to insist that all cables be laid underground.

Thus, although the Exeter Electric Light Company had, in fact, forced the council's hand, their ploy was foiled by the council. As a result, by 1892 all the overhead cables had been placed underground, although this was at a huge cost to the Electric Light Company, at a time when they could ill afford such expense. This unforeseen outlay was but one of the many financial, and other, problems that plagued the entire existence of the Exeter Electric Light Company. Nothing, it seemed, could go according to the plans the directors had made. There were other problems with the machinery installed at the Rockfield Works, although these were quickly sorted and the business stuttered on.

At the 1890 Annual General Meeting of the Exeter Electric Light Company, the auditors could only show a profit of £165.6s.7d. By 1892 this had turned to a loss of just under £14, although the company obviously looked ahead optimistically with high hopes, for another Babcock and Wilcox boiler was purchased, together with another Fowler engine and a Brush-Morley alternator!

At the outset, the vast majority of properties into which electricity was taken were commercial undertakings rather than domestic. One of the first "customers" to receive electricity from the new company was Rockfield's near-neighbour, Exeter's *Theatre Royal*, at the junction of New North Road and Longbrook Street.

The Theatre had re-opened in October 1889 after the original building on that site had been devastated by fire in 1887. At the Coroner's Inquest, the public were told of how the fire was started by the naked-flame "fish-tail" gas lighting system on the stage setting alight a backcloth. It was commonplace in those days, of course, for theatres to be lit by gas, and "fish-tail" burners were readily accepted as a suitable way to light a stage.

An artist's impression of the Theatre Royal fire in 1887.

As a result, there followed what is acknowledged to be the worst fire in the history of English theatre, when over one hundred and eighty people lost their lives. To put that more into perspective, just over three hundred people died in air raids on Exeter during the Second World War.

Exeter's Theatre Royal was re-built, and the second theatre on the site was one of the first theatres in England to be lit totally by electricity, although improved gas lighting had also been installed in the building as a backup in the event of an emergency, or power failure. At the re-opening of Exeter's Theatre Royal, just two years later, the Directors took great pride in proclaiming that:

'....*the building was now illuminated by no less than 510 electric lamps....*'.

The directors had taken quite a bold step by accepting the architect's lighting design using electricity, and the Theatre Royal was certainly one of the first buildings in Exeter to be "lit up" by this revolutionary method.

Note: Electric lighting had been installed at The Savoy Theatre, in London, as early as 1882, and this was possibly the first theatre to be lit by electricity in England. That building was demolished in 1929, and another built on the site.

The Theatre Royal, Exeter.

Seen here at the beginning of the twentieth century, the Theatre Royal was one of the first buildings in Exeter to be lit by electricity.

For some years all went well for the Exeter Electric Light Company, but in 1896 it was once again suffering cash-flow problems, and faced a dilemma. The reason for the dilemma was that demand for electricity was increasing almost by the day, and it was becoming apparent that, in a very short time, the Rockfield premises would be unable to cope with that demand. Massingham's company needed to increase business to meet its customers' requirements, but the building it occupied was not physically capable of being extended in order to house more equipment, and in any case, the company's financial position at the time made it impossible to contemplate any move to larger premises.

More cash needed to be injected, but the company could not offer further debentures to shareholders, as that was restricted by law to the amount raised by Ordinary shares, and not enough had been taken up. Thus the company was unable to raise finance to provide the necessary assets to relocate and be capable of increasing its business.

If the company was unable to raise further money from shareholders, it needed private funding – and yet another problem facing Massingham and his partners was that their efforts to raise such further private funding had not been successful. Whilst the necessary money *could* be obtained by way of loans from the Local Government Board, here Massingham was faced with yet another problem. The only body capable of obtaining such finance from the Local Government Board was the local authority – not privately owned companies.

The company was thus forced into a corner. In order to continue operating, the company needed to move to a bigger building, and install additional equipment. To do this required additional finance, and that could not be found. The company could not apply to The Local Government Board for loans, and as Massingham and his colleagues were unable to secure more private investment, or seek more help from shareholders, the company seemed unable to move forward, as was undoubtedly essential.

Eventually Massingham and his colleagues realised that their only way out was, in fact, to sell the business, and they set about seeking offers for their company as a going concern. Various offers from other companies were quickly received, but none proved acceptable, largely due to the conditions involving debenture holders being required to give up certain rights, such as waiving priority of payment, or transferring their holdings to nominees of the purchasing company.

When Exeter City Council became aware of the situation, it, too, was faced with a problem – for it realised that whilst the future of electricity in Exeter was undoubtedly assured, the city desperately needed a company capable of supplying an ever-increasing demand for electricity should the Exeter Electric Light Company go out of business – and that certainly seemed to be virtually inevitable in the circumstances. Some of the shareholders of Exeter Electric Light Company had argued that the company was only worth in the region of £7,000, although others were in favour of seeking a considerably higher sum. Eventually it was agreed to offer the company to Exeter City Council for £10,000.

This figure was not acceptable to the city council, and in a report by Donald Cameron, the City Surveyor, recommended it should offer just £7,000. Cameron also recommended that the Council should not act solely on his estimation of the value of the Electric Light Company, but suggested an independent assessment be made. The council accepted this course of action, and retained the services of Dr J.A. Fleming, a Professor of Electrical Engineering at University College, London, to advise on the value of the company. This was to prove a wise and shrewd move, for Fleming was also a consultant engineer to the City of London Electric Lighting Company, and thus had gained considerable knowledge in this field.

Fleming travelled to Devon and carried out a lengthy investigation of the Exeter undertaking, of which he proved to be highly critical. He firstly considered the buildings unsuitable for their new purpose, mainly because although they were ideally located next to a railway station, they were a long way from any natural water supply that could provide adequate condensation for the operation it was expected to carry out. Additionally, there was literally no room for further expansion at the New North Road site, and expansion was undoubtedly required.

Fleming was also concerned that the pipework in the original layout at The Rockfield Works had not been duplicated; his argument being that this could cause a possible risk of non-supply should any of the pipes fail for any reason. The risk the company would face should such a failure take place was, in his opinion, too great to contemplate. Failure to supply, even for a very short time, could turn away hundreds of existing and potential customers. He considered the transformers too old, too small and thus inefficient, calculating that for every unit of energy supplied, one was being wasted because of their inefficiency. He concluded that all the transformers were nothing more than worthless, and suggested that if the company was to continue in business it should "get rid of them all".

The boiler room Fleming considered too limited in size, for it would only take one more boiler, probably not enough for the future requirements. With regard to the existing cables supplying the city, Fleming was quite concerned that they would limit further supply to a mere two thousand lamps – and because of increasing demand that, in his view, was far from enough.

Fleming's highly critical report was presented to the city councillors along with that of Donald Cameron. Both men had the expertise to detail the faults at the Rockfield Works, which made them come to the conclusion that the value of the Exeter Electric Light Company, as it stood, was no more than £7,500.

The council readily accepted their conclusions, and this figure was put to Massingham and his partners, who reluctantly were forced to agree to the figure, as they could not dispute those faults brought to light by Fleming and Cameron. The company had little in the way of alternatives to offer its shareholders, and as a result, the council's offer of £7,500 was formally accepted, and the sale went through.

After the sale had been agreed, in September of 1895 the city council approached the Local Government Board to borrow the £7,500 for the works, and for a further £3,500 for the necessary additions and improvements. It was deemed necessary for this application to be made subject of a public enquiry, and this was duly held at Exeter Guildhall in January of the following year. Objections were raised to the proposals, particularly from the city's ratepayers who opposed the burden of the debt that would follow as a result of the purchase – a debt that would, of course, fall back on those ratepayers at some stage.

It is interesting to note that, although Cameron and Fleming had agreed a figure of £7,500 as a value of the Exeter Electric Light Company, at the public enquiry Hector Munro, the City Electrical Engineer, actually valued the undertaking at £12,000! Exactly how he came to that figure is not known, but it put the company's worth at £1,000 more than the sum being applied for! Obviously he was persuasive in his arguments, for in March 1896, following the enquiry, the Local Government Board agreed to sanction a loan of £7,500, and in the following May a further £3,500 was sanctioned.

Formal notice of the sale in the sum of £7,500 was then issued to the Exeter Electric Light Company, and Exeter City Council took over the operation, re-naming it The City of Exeter Electric Company – still at that time operating from the Rockfield Works.

Following the sale, Massingham set up an electrical consultancy business based in Exeter and Torquay. This, however, did not work out in the way he anticipated, and Massingham was declared bankrupt in 1901.

Following a spell of ill health, Henry Massingham moved to Brighton and opted out of the electrical business altogether, preferring instead to run a small hotel.

The Public Health Act, 1875,

AND

The Electric Lighting Act, 1882.

EXETER.

WHEREAS the Town Council of the City of Exeter have applied to the Local Government Board for sanction to borrow £11,000 for the purchase of the Undertaking of the Exeter Electric Lighting Company, Limited, and for the construction of Works of Electric Lighting in the event of the said Town Council being empowered to purchase the said Undertaking in pursuance of Art. 68 (b) of the Exeter Electric Lighting Order, 1891 :

AND WHEREAS the Local Government Board have directed Inquiry into the subject-matter of such Application :

NOTICE IS HEREBY GIVEN that Colonel John Ord Hasted, R.E , the Inspector appointed to hold the said Inquiry, will attend for that purpose at the Guildhall, Exeter, on Thursday, the Sixteenth day of January, 1896, at Half-past Eleven o'clock in the Forenoon, and will then and there be prepared to receive the evidence of any persons interested in the matter of the said Inquiry.

HUGH OWEN

Secretary.

Local Government Board,

1st January, 1896.

Printed by WatErlow Bros. & Layton, Limited, London.   58089—(100)—1-1 96

Notice of a Public Enquiry.

The enquiry examined the city council's proposals to borrow £11,000 to enable the purchase of Exeter Electric Lighting Company Ltd.

Hector Munro Esq., M.Inst.E.E.

Chief Electrical Engineer to Exeter City Council during the planning
and construction period of the Haven Road power station.

Photograph courtesy SWEHS.

THE CITY OF EXETER ELECTRIC COMPANY

Massingham's company had succeeded in bringing electricity to Exeter, although, as has been seen, financial problems made it impossible for his company to continue its operations despite the fact that there was obviously a huge future for electricity, not only in Exeter, but worldwide – perhaps more than anyone could have visualised at that time.

After the city council took over the supply of Exeter's electricity, its main priority was to install new plant to meet the anticipated increase in demand. In 1896 new equipment was ordered, but the need to have this ready for the winter of 1896/1897 never materialised due to various delays. It finally came about in July 1897, six months late.

By this time Donald Cameron had devised a system of lighting the main street from Exe Bridge to St Anne's Chapel in Sidwell Street. This was also fraught with delays because of the need to supply electricity that was over and above the capacity of some of the existing machinery. For example, some twenty-two lamps needed to be changed from D.C to A.C., in order that one machine could cope with the output the lamps required, and this changeover was at a cost of £1,650. New posts and brackets were also required, the contract being given to Francis Parkin of Exeter. Crompton & Company was given the contract for the same number of lamps, and all of this alone totalled a further £400. Again, delays in delivery prevented the changeover happening on schedule, and it was a further twelve months before all of the posts and lamps were in position.

The council, proud of its improved lighting scheme, gave notice that it would come into operation in June 1898. Henry Massingham, however, must have smiled to himself when it was the city council's turn to have problems. Very soon after the new system started, the lamps proved faulty. Some failed because they were not designed to work on the 2Kv. I.I.T. circuits that had been installed, and others because of failing mechanisms. The council quickly ordered new lamps, and these were installed by the November of that year. More problems arose when it was found that the company supplying the switches originally ordered could not deliver – and whilst this was quickly overcome, there were further problems with the replacement lamps! The necessary repairs and replacements meant that by July of 1899 Cameron's grand lighting scheme for the city centre had still not achieved full operational success, but eventually, of course, all the problems were resolved.

Over this period, the demand for electricity had increased enormously, and during the early part of 1899 Hector Munro became concerned about the ability of the council to meet the even higher demand expected during the forthcoming winter. He made his concerns known to the city council, and once again much of the equipment was replaced to increase output. All of this, of course, was at high cost and such costs would eventually, and inevitably, fall back on the consumers.

At the time of the sale to the city council, the Rockfield Works was producing an output capacity of some 200kw. Within a very short time, improvements and better equipment increased this to 321kw. By 1898, Munro's concern that the Rockfield Works was no longer capable of supplying the increasing amount of electricity that would be required was proving to be more than accurate – as, of course, Massingham had realised several years before. By 1899 the output had grown to 575kw, nearly three times the output capacity at the time of the sale. It was time for some serious thinking with regard to new premises.

Hector Munro's concern that Rockfield was no longer a viable proposition prompted him to urge the city council to look at possible sites suitable for a new, and much larger, generating station. Several schemes and proposals were put forward, including sites at a former mill in Countess Wear *(see details later)*, another mill at Cricklepit (on Exeter's Quayside), Belle Isle, off Topsham Road (at that time the city's sewerage works) and land at Exe Bridge. In addition, the Council also looked further into the possibility of trying to increase the size of the Rockfield Works – which was to prove totally impossible, as Fleming had previously advised the council in his report.

All of these sites were given due consideration, including even the possibility of creating a hydro-electric system operating from Trew's Weir. As has been seen earlier, this was something that Massingham had planned, although he, of course, had seen those plans rejected by the very council that was now considering the same proposal! Whilst the idea of using water power via Trew's Weir was given serious thought, it was considered the proposal could possibly be detrimental to Exeter's then busy canal, and the idea was taken no further.

In 1899, a site in Haven Road was purchased for £1,000, and later the same year another potential site, the redundant paper mills at Countess Wear, a suburb of Exeter, was also purchased for £5,000. Both sites, of course, were close to a substantial water supply, but Countess Wear had no immediate access to a railway, although at first that was not considered a serious problem, as coal could be transported by river. Later consideration proved otherwise.

Eventually Munro was able to persuade the council that their best option was to build a new power station on the vacant land that they had purchased adjacent to the City Basin, in Haven Road, a site that was reasonably close to the city centre, adjacent to the Exeter Canal, and with a rail system already in place that literally passed the door. Since the land had already been acquired by the council, it was a fairly simple process to carry the idea forward, and plans were drawn up for a new Exeter power station.

However, before taking any final decision, the city council decided to employ another expert, T.E. Wilmshurst Esq., MIEE, the Borough Electrical Engineer at Derby, in the capacity of an advisor. Wilmshurst was another person who enjoyed a considerable reputation in the industry, and it transpired that his endorsement of the Haven Road site above any of the other possibilities was the key factor that convinced the city council to go ahead.

The Exeter Basin, 1911 – an early postcard.

An important part of Exeter's trade and industry for many years, particularly the woollen trade in the seventeenth and eighteenth centuries. Until fairly recently, The Basin saw frequent visits by oil tankers, timber boats and other cargo vessels. It is still very much in use every day – although now mainly for recreational purposes.

After the Basin was constructed, it was also known as *The Haven* – the word Haven being defined as 'a sheltered place for shipping'. Hence the names Haven Road and Haven Banks.

At the beginning of 1901, having finally chosen the appropriate location for the new power station, the city council invited tenders for the construction of the new building, together with the necessary generating equipment required and the large amount of ancillary machinery that would also be necessary to make the power station a success.

Council minutes report that the specifications for the new plant included a capacity of some 1,250kw at 2,000 volts, with the boilers being required to be capable of up to 160lbs per square inch pressure, and able to supply 52,000lbs of steam every hour. It was decided that system of supply would be two-phase, particularly as this would involve less expense.

Despite Hector Munro and others opting for a tender from C.A. Parsons & Co., a company that had installed similar equipment elsewhere in the country, the council decided to accept the tender of British Westinghouse, even though it was slightly higher than that of Parsons. In fact, the Westinghouse tender of £29,707 was increased later, due to the addition of chain-grate mechanical stokers (£1,486), and also the inclusion of high-tension feeders to sub-stations in South Street and High Street (a further £2,450), making their total tender £33,643.

As is common in such schemes, having won the contract with Exeter City Council, Westinghouse sub-contracted much of the work involved, and invited tenders from other companies to cover several aspects of the scheme. Westinghouse duly appointed several experienced companies to carry out various aspects of the main contract.

Some of these were well-known national companies, and two were awarded to local companies – Messrs Willey and Co. Ltd, whose iron foundry was at Water Lane, within a few yards of the Haven Road site. The main construction work was awarded to an Exeter building firm, William Brealy.

The companies submitting successful tenders were as follows:

- Messrs Babcock & Wilcox – the supply of boilers each capable of evaporating 13,000lbs of water every hour.

- Messrs Bellis & Morcom – the supply of three 570bhp triple expansion engines linked to 400kw alternators to be supplied by Westinghouse.

- Worthington Pumping Engine Co. – supply of the condensers.

- Messrs E. Green & Co. – supply of the economisers.

- Messrs Willey & Co (Exeter) – the supply of a coal-converting plant, plus ironwork for the buildings, coal bunkers, and pipework, etc.

- Messrs C. & A. Musker – supply of a 10-ton travelling overhead crane.

- William Brealy, a local building contractor with premises in nearby Cowick Street, was awarded the contract for constructing the building itself in the sum of £13,905.

- Westinghouse provided three of its own generators, each being 400kw (of 20 poles), and another being 100kw (of 14 poles), and all designed to generate at 2,200 volts at 60 cycles per second. They also supplied the alternators mentioned above.

Building work was scheduled to commence at the beginning of 1902, and later on, during the period 1903 and 1904, Henry Massingham was, perhaps somewhat surprisingly, employed to supervise the installation of the electrical and steam equipment, under the ever-watchful eye of Hector Munro. Munro was, of course, well aware that Massingham's original company had experienced financial difficulties under his ownership, and that his subsequent electrical contracting business in Exeter had ended up in receivership. He was therefore not totally convinced that Massingham was the ideal man to oversee the work without constant supervision!

The location of the new Power Station in Haven Road.

Seen here as the large building in the centre of the map, with the two chimneys. The Basin Branch line comes up from the bottom centre, with a siding running from it to the rear of the Power Station.

The Great Western *Basin Branch* came in from the main line that runs from Exeter St David's station to the South Devon coast.
Only fragments of the branch line now exist.

COUNTESS WEAR MILL

One of the possible locations to be given serious consideration by the city council for a new power station, was the former paper mill at Countess Wear, then but a village on the edge of Exeter. The mill was at that time in partial ruin and had not been used for many years. It is worth recording a little of the history of these premises within this story.

Whilst Countess Wear is now a sizeable area of Exeter, it was originally a small village, and a part of Topsham. It is, perhaps, best known locally for the thirteenth century weir across the River Exe constructed by Isabella, the Countess of Devon, allegedly to power her mills further upstream. Various stories, many of which differ, have been written about the problems caused by this weir, including the fact that it destroyed Exeter's salmon fishing on the Exe, prevented shipping reaching the important port of Exeter, and gave to the grievance that the weir allowed more trade to the lower port of Topsham – controlled in those days by the Earls of Devon from their seat at nearby Powderham. The exact truth, however, is a somewhat grey area that warrants no further investigation here.

The village of Countess Wear was a thriving community. There were lime-kilns, boat-builders, glass-makers and a flourishing paper mill, all of which were in addition to the various other occupations of the day, particularly agriculture. The paper mill was constructed during the seventeenth century, on the site of an original grist mill. The mill was destroyed, on at least two occasions, by fire, an occurrence that was not uncommon in industrial buildings in the past. Each time it was rebuilt and continued to operate as a successful business until 1885, when it finally closed.

Whilst much of the original building was then demolished, those buildings that remained, together with the surrounding land, were considered by the city council as a possible site for their proposed sewerage farm, but they instead opted for a site further up the river, at Belle Isle, a short distance downstream from Trew's Weir.

Countess Wear Mill, 1940.

The former mill in Countess Wear village was one of the possible locations for the new power station. For various reasons it was rejected.

The mill was then put forward as a possible location for the proposed power station, but one of the factors against this proposal was that it would be necessary for coal to be brought to Exeter from elsewhere in the country, and it was far easier to do this by rail rather than by sea. However, although it would not have been impossible to construct a new branch line from the existing main line to the Countess Wear site, it would have necessitated crossing both the ship canal and the River Exe. The cost of carrying out this work meant that to locate the power station at Countess Wear would have made the project financially inviable. To transport the coal by sea, and eventually up the River Exe to Countess Wear was also dismissed as far too costly, mainly due to the fact that whilst many of the country's coal mines were close to existing rail networks, few were near the coast, necessitating further transport costs before the coal could be loaded at a port.

Thus, Countess Wear Mill was dismissed as an option for the new power station. Instead, the site was purchased by local resident Percy Sladen.

Walter Percy Sladen was a marine biologist who specialised in the study of asteroidea (starfish) and other marine animals known as echinoderms. Sladen, and his wife Constance, lived latterly at Northbrook Park, one of Exeter's many out-of-town residences occupied by wealthy gentlemen of that time. Northbrook stood in the considerable grounds now occupied by both the Exeter and Devon Crematorium and the Northbrook pitch-and-putt golf course.

Sladen had inherited the Northbrook estate on the death of his uncle, John Dawson, in 1894. Sladen and his wife, Constance, moved in to their new home soon afterwards but unfortunately, Sladen's health was deteriorating at that stage, and he only lived there for two years, for he died in June of 1900 whilst on yet another of his scientific trips abroad.

Somewhat ironically, perhaps, Sladen had purchased Countess Wear Mill in 1899, solely to install the necessary machinery to supply his house with electricity, and during the Second World War this was to be something of a life-line for Exeter. The city council commandeered the property, and it was taken over by the city's Civil Defence. The logical thinking behind this action was more than justified when Exeter was bombed during 1941 and 1942. Several areas of the city, particularly the city centre, lost much of their power for long periods, sometimes for several days. Having taken over a property that enjoyed a private electricity supply, located in an area that was unlikely to be bombed, the Civil Defence was able to continue its work in protecting Exeter, despite Hitler's efforts to destroy the city.

Thus Percy Sladen had unwittingly proved to the council that the mill could be used to provide electricity, certainly on a small scale, if not as a commercial enterprise, although by that time the "new" power station in Haven Road had been operating for almost forty years.

Note: The Royal Albert Memorial Museum in Exeter's Queen Street holds Sladen's large collection of his lifetime work with marine life, including many fine specimens and a library of scientific books.

Walter Percy Sladen.
1849-1900

Reproduced by permission of The Linnean Society of London

THE NEW POWER STATION

In 1903 the foundation stone of the City of Exeter Electric Company's new building was officially laid by Edward. J. Domville Esq., JP, the chairman of the council's Electric Lighting Committee. The architect was Donald Cameron (the city surveyor), the electrical engineer was Hector D. Munro (the city's own electrical engineer), and the construction work carried out by William Brealy of Exeter.

Once the building was finished, it would be possible for the company to supply electricity to a far greater area of the city than the Rockfield Works would ever have been capable of supplying. At the same time, Exeter would be able to run an electric tramway system that would cover the whole of the city centre.

Although it could be seen from other areas where electric tramway systems had already been installed (and from those that were in the process of doing so) that this was the way ahead, when Exeter's power station opened, it was not really anticipated that within a very short time the popularity of the new electric trams would necessitate the tramways operation to be extended, covering some of the outlying areas, such as Whipton and Alphington. This, of course would mean a considerable increase in the amount of electricity required for that operation alone, over and above the already increasing amounts being supplied to businesses and houses in the city.

The changeover from the Rockfield Works to Haven Road took place on 2nd April 1904. In the wake of difficulties experienced with any possible expansion at the Rockfield Works, the design of the new building not only took into consideration the current requirement, it wisely allowed for future expansion of the output from the power station. As electricity gained in popularity, so the demand for it grew almost by the day.

The opening day of the Haven Road Power Station in 1905.

A group of dignitaries and guests pose for the photographer around one of the generators in the Engine House.

Photograph courtesy SWEHS.

As can be seen on the plan on the following page, the site chosen at Haven Road was sufficiently large enough to accommodate additional buildings in the future, and indeed this was necessary a few years later when Heavitree was added to the Exeter network. Much later again, in 1926, nearby Topsham was added, for until then a simple, belt-driven DC generator had struggled to provide Topsham's total supply! Following Topsham, in 1932 the nearby villages of Kenn, Dunchideock and Shillingford St George were also included in the supply.

The power station originally occupied an area of some 150 feet square, embracing the boiler house, engine house, and offices, with annexed buildings for economisers, boiler feed pumps, etc. A photograph in the souvenir brochure of the official opening shows that the building was constructed as three linked buildings, with the gabled roofs of each clearly visible.

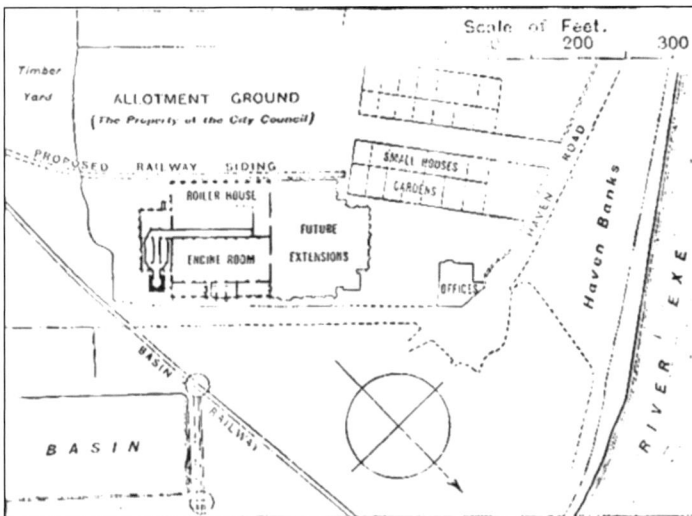

A draft site plan for the proposed Power Station.

This plan, drawn up by architect Donald Cameron in 1903, is not very clear. The coal bunker house is not shown. but the area for "future extensions" can be seen clearly marked to the right of the engine room and boiler house. This area proved essential when the building was extended in the 1920s.

Photograph courtesy SWEHS

The main entrance, in Haven Road, was in the centre of the original symmetrical façade. Large double doors were at the top of granite steps, and these doors were set into a Ham stone porch, comprising a large round-headed hood and moulded soffit, supported by Tuscan columns. Within the porch there is also a Ham stone arch, with an ovolo-moulded surround. The central section is set forward slightly from the main façade, neatly breaking up what would otherwise be a completely flat front. This section is also one storey higher, and above the second-floor windows there are two Ham stone panels, each containing bas-relief sculptures of reclining ladies in Greek-style dress. One is holding an electric light bulb, and the other reading a book. At the very apex of the central section, above the middle window, is another Ham stone panel, and this contains the City of Exeter coat of arms.

Note: To the untrained eye, the stone surrounds could easily be mistaken for Portland stone. Ham stone is from an area of Somerset, more yellow and honeycombed in appearance than Portland or Beer stone.

The central section of the Power Station façade.

The Ham stone panels include Grecian-style figures and the Exeter Coat of Arms
One of the figures, quite appropriately, is holding an electric light bulb and cable.

The main entrance of the Power Station in 2008.

The large, double doors are set in an attractive round-headed arch in Ham stone.

In a report by prepared jointly by Exeter Archaeology and Keystone Historic Buildings Consultants (published 2000), the power station building is described as being *in an eclectic classical style with baroque details, not unusual for Edwardian industrial buildings: it is essentially a version of a Queen Anne mannerist style.* Elsewhere in that report it is described as "imposing", "of architectural presence", and also being "of architectural interest, historical interest and has group value".

Minor details, such as the rainwater goods – the gutters, downpipes, hoppers and even their fixings – were all designed by Cameron, and these are not just bland, uninteresting accoutrements, but pleasant and slightly ornate items that only add to the building's appearance.

Seen from the east side of the River Exe, the building is certainly one of Exeter's attractions, although like some other buildings in Exeter, it is not generally appreciated until pointed out – possibly in this case because it was simply a power station, and some distance from the city centre.

External architectural detail of the building.

The Haven Road building was designed to be attractive, and it certainly contained architectural detail that was not common in buildings designed as power stations at that time. Old photographs show that many other power stations were comparatively bland, basic structures, often-rectangular "blocks", with very little in the way of interest in their design.

Peter Lamb, secretary of the South West Electricity Historical Society, explains that if a private company had been charged with building the property, it would have been a much more utilitarian, and thus cheaper, building, with scant architectural merit. The city council was fortunate in having three main advantage points over private undertakings: (a) it already owned the land; (b) it used its own architect; and (c) it was able to borrow money from Government funds quite easily, whereas private companies would be required to rely on cash advances from shareholders.

The latter in particular could have proved difficult for some undertakings, as it is quite probable that those who were already shareholders would have reacted adversely to investing more of their money for any project in which they had already invested, due to the vast expenditure involved in setting up adequate distribution systems and other such necessary works. It must be remembered that electricity was still, at that time, a comparatively unknown quantity, and something about which few people had any in-depth knowledge. Investors were more concerned about their potential financial return than having electrical knowledge.

When the Exeter City Council chose to use Donald Cameron to design the power station, it was indeed a wise move, for the building is now acknowledged to be one of the finest examples in this country of a power station, designed with few restrictions on the architect. It is still admired by today's architects, electrical historians and Edwardian enthusiasts, and, fortunately, the building will now remain *in situ* for many years to come. Even to the eye of a mere passer-by, with no architectural knowledge, it has a most acceptable appearance.

This newly constructed building was designed in a style that was elaborate without being too ornate, and even the adjacent chimney had some elegance, if that word can suitably describe the appearance of an outlet for smoke! The original chimney – removed during the 1960s – stood one hundred and fifty-five feet high (in excess of fifty metres), and had an internal diameter of eight feet and six inches at the base. The chimney was also considerably decorative in its brickwork, not just a plain stack as many were, and it complemented the building alongside.

The Power Station chimney being built.

Architect Donald Cameron's design included much ornate brickwork, rather than the usual bland stack. Over one hundred and fifty-five feet in height (some fifty metres) and with an internal diameter at the base of eight and a half feet, the chimney was far more ornate in design than most other chimneys in the area. Note the gentleman posing for the camera halfway up a rope to the right of the chimney!
The photograph is thought to be an early photocopy from a journal, or a newspaper, and is thus a poor reproduction.

Photograph courtesy SWEHS.

Exterior photographs taken at the opening in 1905 show the Haven Road façade to be symmetrical. A two-storey section was either side of the central, three-storey section. The flanking sections had three double windows on the ground floor, and a single window either side of a double window on the first floor. The central section was in completely reverse, the ground floor having two single windows either side of the main entrance, with the first floor having three double windows. The second floor of this section had two single windows either side of a slightly larger central window.

The Power Station in 1905, shortly after opening.

The tree trunks are destined for the saw mills that stood between the power station and the Exeter Gas Works. The foreground is now part of the Piazza Terracina.

Today, however, the same façade presents an entirely different view, for the building is almost twice the length of the original. When this extension was carried out is not certain, but research suggests a considerable enlargement, possibly in the late 1920s, to accommodate the additional demand mentioned above. In 1927, the front of the building remained the same size as the original building, as a photograph bearing that date shows, but Ordnance Survey maps show the extended building during the 1930s, and this seems to date the extension between 1927 and 1932. Fortunately, the design of the extension virtually mirrors that of the original building, although sandstone has been used for the window surrounds and cills, etc., which is a slightly different colour to Ham stone, and some surrounds are plain with no decoration. The extended part of the original building is still visible by a distinct line in the roof, where asbestos sheeting has replaced what would probably have been slates.

Note: If the extended part of the building is studied closely, there is certainly reason to suggest that there may well have been even a second extension, possibly carried out during the 1970s when the present electricity sub-station was installed. This has not been verified, however, and could prove to be wrong.

The Power Station in 1927.

The date is just visible, written in ink on the bottom left of the photograph. Haven Road has no tarmac, and there are no footpaths. The second chimney looks to be as tall as the original – but this is a trick of the camera as it was, in fact, much shorter. At this time the building has not been extended.

Photograph courtesy of SWEHS.

The Power Station in 1948.

The building has now been extended to the right – and the "join" can clearly be seen in the roof-line, and this is a feature that is still visible today. This view shows the difference in height of the two chimneys. The railway wagons are on the GWR Basin Branch line.

Exeter could now be proud of its new power station, a building that was considerable larger, and far more efficient, than its predecessor at New North Road. Within two years of the Haven Road premises opening, some four hundred houses were being supplied in the city, over five hundred businesses, and over fifty other premises including public buildings and churches. In 1927, the power output from the Haven Road building had been something in excess of 4000 kilowatts. Just three years later, the demand for electricity was such that the output had increased dramatically to over 15,000 kilowatts, with almost eighteen thousand customers registering with the company, and by the mid-1930s the company was supplying almost twenty thousand customers, including the smaller Exe Valley and East Devon Electric Companies.

Even so, the design of the new facilities at Haven Road ensured that the output capacity was by no means working at maximum. Because there was a built-in allowance for increasing capacity over future years, the company could offer suitably priced electricity, and if necessary alter their prices to suit the demand. The company even employed a "canvasser" who went around the city touting trade, and whilst his basic wage was minimal, he earned commission on whatever contracts he secured.

For example, the canvasser could earn threepence per lamp on the first three thousand lamps he "signed up", and sixpence per lamp on every lamp over that figure. If he was able to sell motors, then the commission was set at five shillings per horse-power, and thus a 10hp motor would earn him fifty shillings. This system of having a salesman seemed to be sensible, and in the first year the canvasser managed to gain eighty-one new customers in just over five thousand calls, earning himself £100 for his trouble – and that was reasonable money in the 1800s.

Note: In those days, two hundred and forty pence comprised a pound. There were twelve pence to every shilling, and twenty shillings to every pound. Two shillings was a tenth of a pound, and thus the equivalent of today's ten pence coin.

At this time, several new properties and housing estates were being constructed in Exeter, and locations such as Newtown, Pennsylvania, Longbrook Street, Heavitree Road all required connection to the new system. There followed further demand in areas such as Cowick Street, Alphington Road, Sidwell Street, Southernhay and several other places where land was being purchased for development.

Nothing would stop the constant demand for electricity, not only in Exeter, but also throughout the country, and indeed, the world. Even so, until comparatively recently there were still a significant number of properties that were without mains electricity, some folk preferring to remain on the old system of private generators, and some simply refusing the "new phenomenon" as they did not trust it!

Today, of course, there are probably but a handful of domestic residences throughout the country without electricity, virtually all of those being in outlying and remote areas. There is probably no business today capable of operating without electricity, especially in this age of the computer, and similar technological advances. Exeter's demand increased all the time, although, of course in some of the more remote areas around the city there was no supply in the early days. That would be rectified several years later when the National Grid assisted in bringing supplies to rural areas.

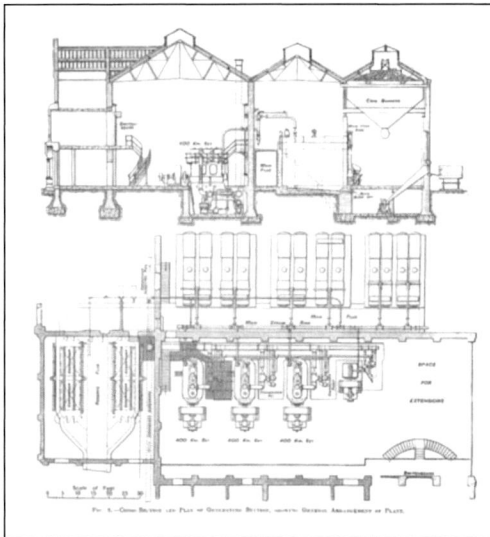

The architect's drawing for the Power Station layout.

Top: The Engine House, Boiler House and Coal Bunker House
Bottom: The Engine House layout, showing the generators linked to the boilers.

Photograph courtesy SWEHS

In 1902, Exeter City Council took over the running of the city's tramways, still horse-drawn at that time. Even at that early stage, consideration was being given to changing the tramway operation to the new electrically-operated vehicles that other operators had introduced in various parts of the country. If this happened, it would, of course, put a huge demand on the power station output.

Hector Munro put it to the council that their best course of action would be to "obtain a 200kw generator and engine set, and a rotary converter as a reserve, to be powered by one of the existing 400kw alternators". This would be sufficient to operate electric trams. He estimated that this would cost in the region of £4,000. The council, however, wanted to use a turbine system, but Munro – once having advocated the use of turbines – now argued that there was little benefit in using turbine power on motors under 500kw. Instead, he put forward an alternative scheme whereby two alternators would be used during the evening, and one kept solely for daytime demand. He suggested that by employing a 200kw DC generator and a converter, the third 400kw alternator could be kept in reserve, to be used when demand was such that more power was needed. He pointed out that this would be particularly relevant in winter evenings, when the demand was high and surges in power frequent.

The council eventually agreed to his plan, and Westinghouse was asked to supply a new 200kw dynamo, coupled to a Bellis & Morcom engine, all for the sum of £2,260. Bruce-Peebles supplied a 200kw rotary converter at £1,050 and this machinery was installed in towards the end of 1905, although it proved to be unreliable and faulty, and in 1906 a replacement was provided by Bruce-Peebles at their expense.

By early 1905, all the work covered by the original contracts to construct the power station was completed. Within two years, with all the faults and teething-problems sorted out, the City of Exeter Electric Company was supplying electric light and power to domestic dwellings, businesses, public buildings and places of worship throughout the city. From Massingham's early experiments, electricity in Exeter, the company were now able to enjoy an incredible increase in demand. The future of this new service was assured, and over the next half-century, the power station would prove a successful investment by the council.

Electricity was undoubtedly here to stay!

THE ENGINE HOUSE & OFFICES

The building that can be seen in Haven Road today comprised the former engine room and ancillary offices, and is just a third of the original structure. The engine room itself is one hundred and two feet in length, and some forty-five feet wide, and is unusual in that it is lined completely with white glazed bricks, occasionally enhanced with decorative beige-coloured stone and brickwork. Tall, arched recesses make the long walls far more interesting, although the arches do not appear to have served any other purpose. Originally, the floor was of "terrazzo" marble paving, which unfortunately no longer exists. The reason given for the glazed bricks and marble floor was "to ensure the utmost cleanliness". Even some of the doorways and openings were arched, again making them more interesting and easy on the eye.

Along the whole length of one side is a gallery, with an ornate cast iron balustrade. Decorative posts every few feet are still in position, and originally they were topped with circular electric light shades – although sadly these are now missing, but hopefully they will be replaced in the course of renovation.

The gallery is supported by several cast iron wall brackets, the casting of each being designed to include the "three castles" from the Exeter coat of arms. The gallery could be accessed from the first-floor offices, and gave an uninterrupted view of the whole of the engine room. There were originally two cast iron staircases, from the engine house floor to the gallery, although only one is still in place today. Another, leading to the boiler house, has also been removed. A stamp included in the casting shows that these staircases were from the foundry of J. Allasen & Son, in Glasgow.

The new Engine House.

This photograph is shown in the souvenir brochure distributed to guests at the opening day in 1905.

High above the gallery, rails are built along the whole length of both side walls, enabling the overhead cranes that span the entire width of the engine house to travel from one end of the building to the other. The rails and the overhead cranes were by engineers Babcock and Wilcox (London & Renfrew), and are still in position. It is interesting to note that although the souvenir brochure issued at the opening of the power station states that the cranes were capable of lifting machinery of up to five tons in weight, one crane still in position clearly displays a safe lifting weight of up to twenty-five tons, and the other up to five tons, so whether these are the original cranes is not known. The brochure details may be in error, or possibly one (or perhaps both) of the cranes were replaced at some stage. As far as is known, there are no records in existence to prove which is correct.

Engine House generators

The photograph above was taken in 1925, and shows the Gas Works end of the engine house. The date of the photograph below is unknown, but was possibly taken around the same time.

Photographs courtesy SWEHS.

The original equipment in the power station was eventually replaced.

In these photographs, modern alternators are being serviced in 1955.

Photographs courtesy SWEHS.

In one corner on the floor of the engine house can still be seen a Hewittic rectifier cabinet, type 1/100/3, manufactured by *Hackbridge and Hewittic* of Walton-on-Thames, Surrey. It is thought that this rectifier would have supplied the current for the overhead cranes.

The identification plate on the cabinet door states that the rectifier had a three-phase AC supply of 415v, 50 cycles, and a direct current output of 525 volts, 34.3 amps. Unfortunately this is now just a cabinet, the workings inside having been removed.

The Hewittic rectifier cabinet.

The name plate still fixed to the rectifier cabinet door.

Also on the floor of the engine room were five steam alternators, having a total capacity of 1,300 kilowatts, capable of supplying fifty thousand lamps or their equivalent. One 200-watt steam-driven generator was provided solely for the Exeter tramway system, essential after the old horse trams were replaced by electric trams in 1905. There was also a motor generator of equal capacity, taking current from the alternators and converting it into current suitable for the tramways.

For the technically minded, the souvenir brochure of the opening of the power station states that each steam alternator consisted of a high-speed vertical steam engine, with forced oil lubrication, direct-coupled to a two-phase Westinghouse alternator and exciter. Jet condensers in the basement of the building dealt with the exhaust steam leaving the engines, and these drew their cooling water from a large well outside of the building, which was connected with the City Basin. Surplus water was returned to the Basin after being used.

The original engines ran at a modest 150lb per square inch of steam pressure, driving the two-phase alternating current generators. Later, in the 1920s, five turbines replaced the original plant, and these ran on a three-phase system, which was at that time a much more accepted method than the original. The turbines remained in the building until it closed in 1960.

The current from the tramway alternators and generators was taken by lead-covered cables in stoneware conduits beneath the floor of the building to the switchboards on the main gallery, and distributed throughout the city by main feeder cables – also lead covered, and again laid in stoneware conduits. Some of these conduits were actually discovered within the power station during the renovation work.

A section of one of the exposed conduits.

The lighting switchboard consisted of marble panels for each alternator, and for each section of the main feeder cables, all fitted with all the necessary switches, indicators, regulators and recording instruments. A separate switchboard controlled the tramway generators and feeder cables. The switchboard took up a large area in the power station, and was huge in comparison to today's compact, digital units.

Front (above) and rear views of the main switchboard.

Photographs courtesy SWEHS

The 5-ton crane in the Engine House.

The 25-ton crane – the largest in the Engine House.

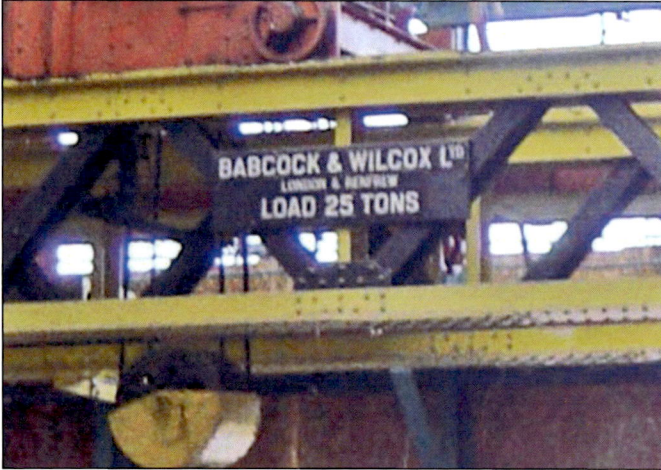

Name plate and safe load sign on the 25-ton crane.

The "railway" on which the Engine House cranes run.

The Engine House gallery in the 1960s.

No longer a power station, but at that time still being used by the
South West Electricity Board.
The gallery balustrade had lamps with circular shades, as shown above,
but these are now missing.

A close-up of the gallery balustrade in 2008.

One of the gallery lamp-posts in 2008, minus the original glass lamp shade.

A gallery support bracket, showing the three castles from the Exeter coat of arms.

The internal glazed brickwork of the Engine House, seen in more detail in the photograph below.

Glazed brickwork helped to ensure cleanliness in the Engine House. Coloured dados and curved arches made it more interesting.

Attached to the engine house, and fronting Haven Road, there was a two-storey building that comprised various offices, storerooms, toilets and mess rooms. A third, smaller, storey was on the second floor over the main entrance.

The power station was entered from Haven Road. The entrance hall and the inner hall floors were originally covered with patterned polychrome tiles. The outer and inner halls are still separated by the original half-glazed door, set into a large, rounded arch with glass panels, moulding and architrave. The main staircase, with stone steps, leads from the entrance vestibule to the offices, and has delightful cast-iron balusters on which rests a hardwood handrail, all curving gracefully around the staircase over three floors. The first step incorporates a cast-iron newel post support. The whole of the staircase is still in position today.

A reeded dado rail stretches the full length of the staircase, only broken at the foot of the main staircase by the slate foundation stone, is set into a carved stone surround that includes the Exeter coat of arms at the top. The plaque has fortunately escaped any serious damage by vandals, and records the fact that the foundation stone was laid by Edward Domville in 1903.

Note: Edward J. Domville OBE., was, for several years, chairman of the city council's lighting committee. He was also the first chairman of the Exeter Theatre Company – from 1887 the body that ran Exeter's Theatre Royal, itself one of the first buildings in Exeter to be lit by electricity. A consultant surgeon at the Royal Devon and Exeter Hospital, Edward Domville was elected Mayor of Exeter in 1893.

The main entrance, seen from the interior in 2008.

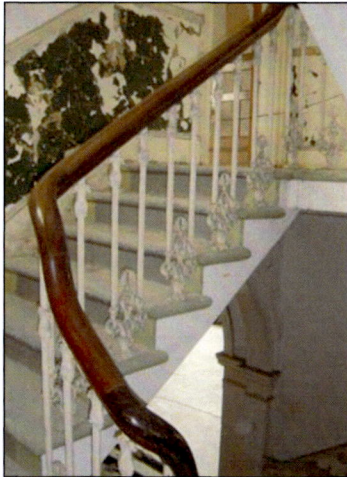

The staircase leading from the vestibule.

The foundation stone in the entrance hall.

The ground-floor rooms are of little architectural interest, and are only accessible from the engine house. On this floor there was a repair shop, fitted with electrically-driven tools – quite an innovation at the time, but of course standard these days. There were also various test rooms, including a cable room where the mains could be tested at any given time.

On the first floor was a general office, together with ancillary other offices and stores. The first-floor rooms have interesting doorways and moulded brick cornices. Most of the Edwardian (or possibly Victorian) fire grates have been removed, or have been vandalised. Some of the original wooden parquet flooring is still visible, and will hopefully be retained. The main office on this level is accessed from the staircase landing, and has an attractive doorway. Two large windows look out on to Haven Road, and these contain stained glass in the transoms. These first-floor offices also had access to the gallery, giving a total view of the engine house.

The second-floor office, originally occupied by the Chief Electrical Engineer, is a pleasantly light and airy room, with a large, arched stained glass window, and smaller single windows. Again, this room overlooks Haven Road and has a good view of the Basin area. From the two end rooms on the second floor, doorways led out to the flat roofs of the first floor rooms. From here can be seen various parts of Exeter, including the Exeter Cathedral, and downstream of the River Exe towards St Leonards. There is also a splendid view to the rear, over St Thomas to the Haldon Hills.

An attractive stained glass window in the stairwell.

Sadly, several other similar windows have been vandalised.

Views from the first floor roof level.

Above, over The Quay towards Colleton Crescent.

Below, across St Thomas to the Haldon Hills.

Some of the new apartments adjoining the Power Station enjoy similar views.

THE BOILER & COAL BUNKER HOUSES

The boiler house and the coal bunker house were separate buildings, although attached to each other. The boiler house, being the central building, was also connected to the engine house. It was essential that all three buildings were linked, as each was complementary to the other. The massive boilers were housed in the central section of the building, but the rear of each protruded slightly into the coal bunker house in order that they could be serviced. Coal was automatically fed down chutes into the furnaces that heated the boilers, the furnaces being under each boiler, and again the rear of the furnaces projected into the coal bunker house. The boilers, of course were linked to the engine house as they supplied the steam to power the generators.

All three buildings were 102 feet in length, but the boiler house was 10 feet wider than the other two, being 55 feet, rather than 45 feet. Within the boiler house there were six water-tube boilers, each fitted with super-heaters, and capable of supplying steam for 2,700 horse-power. A steam-pipe ring on the boiler house wall carried steam through to the adjacent engine room, and had been designed so that any boiler or engine could be isolated without affecting the general supply. The main flue, behind the boilers, carried hot gases through the economizers on the way to the external chimney, thus utilising the surplus heat to warm water pumped into the boilers.

The boilers chosen for Exeter's new power station were designed and constructed by an American company, Babcock and Wilcox. Established in 1867, the company was originally set up in Rhode Island by George Babcock and Stephen Wilcox. Still in existence, the company now supplies boilers in more than ninety countries worldwide. Their early boilers were described as "Non-Explosive"!

As the design included the use of specialist tubing to generate higher pressures, the boilers were more efficient, but at the same time proved safer than other designs. During the Second World War over half of the U.S. fleet was powered by Babcock and Wilcox boilers – quite a testament.

The secondary, metal chimney.

Located at the Diamond Road end, it was connected directly to the boiler house. Some of the houses in Diamond Road are on the right hand side.

Photograph courtesy SWEHS

Coal chutes feeding the furnaces in 1923.

Overhead hoppers were automatically loaded with coal from railway wagons on the branch line alongside the building. The coal was then fed (again automatically) to the furnaces below each boiler when required.

Photograph courtesy SWEHS

A feature of the boiler house was a complete system for the automatic handling of coal and its economical combustion. Regular supplies of coal arrived at the power station by rail, and were delivered into chutes, which conveyed it to the automatic filler of a gravity bucket coal conveyor, which was electrically driven to convey the coal to a series of large bunkers, capable of holding some five hundred tons of coal. Each bunker had a chute, which fed carefully measured amounts of coal into two hoppers, ready for use. The chain-grate mechanical stokers then automatically fed the coal to the furnaces of the boilers, and after use, the ash and clinker was deposited into the basement, from where it was taken – again automatically – by conveyor to the ash bunker outside the building, and eventually removed by being loaded into rail trucks on the railway siding that came to the rear of the building from the Basin Branch Line.

The furnaces, it was claimed, were capable of burning the cheapest of coal and coke without emitting excessive smoke – although even the emissions given off then would not be permissible in today's "smoke-free" city.

The Coal Bunker House in 1927.

Another view showing the coal chutes feeding the furnaces. The circular boilers can be seen in the top left of the photograph. Each boiler had a meter registering the amount of coal used, one of which can be seen as a small, rectangular box immediately to the left of the first chute. It is embossed *Lea Coal Meter*.

Photograph courtesy SWEHS

On occasions it was necessary to release pressure from the boilers, and this was done by opening safety valves, known by employees at the power station as 'blow-off' valves, located under the coal-chute house. When opened, these released a fine stream of hot vapour. Alderman Roy Hill, a past Mayor of Exeter, recalls that when his father worked in the boiler house he was allowed to go into the building occasionally, and it was here that Roy claims to have had his first shower when the blow-off valves were opened!

The rear of the Coal Bunker House – then standing empty.

The railway siding track is just visible. On the right are some of the buildings of neighbouring Claridge's Saw Mill.

Photograph courtesy SWEHS

The Westcountry railway network played a vital part in the running of Exeter's power station. Not only did it bring in coal for the boilers from various parts of the country, it was also used to take away the waste material afterwards.

The Great Western Railway main line to the south Devon coast left St David's Station and headed via St Thomas Station (in Cowick Street) to Dawlish, Teignmouth and onwards. At Willeys Avenue a separate line branched off to Alphington Halt, and linked with the Teign Valley line. The 1932 Ordnance map shows that line as "Exeter Railway". A short distance after Willeys Avenue (at a point approximately to the rear of where the Exeter Recycling Centre now stands in Marsh Barton) another line branched off to the Exeter Basin, and this was named the Basin Branch line. It terminated at the Haven Road side of the river, opposite the Quay, and served both the power station and the Exeter Gas Works, located just a few yards along Haven Road from the power station.

The Basin Branch line was also connected to the Exeter Railway branch, with a separate link going through Tan Lane and past the Basin Junction. A mere fragment of the line remains today.

The Basin Branch Line crossing Haven Road to the Basin.
Behind the corrugated sheeting fence can be seen the power station.
Signs such as that seen above are now sought after by railway enthusiasts.

The restored turntable location at The Basin.

A preserved section of rail track leads away from it – the only section of
The Basin Branch Line remaining.

The plaque alongside the turntable. It reads:

*This turntable, one of two serving the canal Basin, was constructed
to carry standard and broad gauge trucks. The site was excavated and
made a feature of by the Exeter City Council to celebrate GWR 150
24 September 1985.*

The second turntable, exposed in August 2008.

This section will be suitably protected and left underground,
the cost of lifting and renovating it being too prohibitive at present.

The interior of the City Basin Signalbox.

Formerly located on the railway embankment in Haven Road.
Like many others of that era, this box has now been demolished.

Neither the Basin Branch line nor the Exeter Railway branch through Alphington exists today – but the memories do, for the Teign Valley line was "something special" for many local residents. The former Teign Valley line was a popular excursion line. It opened in 1903, at about the same time as the power station construction started. The line ran from St David's Station via St Thomas station, Alphington Halt, and then to Ide Halt, Longdown, Dunsford and Christow. In later years the line was extended to include Ashton, Trusham, Chudleigh, Chudleigh Knighton and Heathfield.

Alphington Halt operated from 1928 until 1958 and neighbouring Ide Halt from 1903 until 1958. Virtually all the remaining stations closed for passenger traffic in 1958, although one or two continued for the carriage of parcels. It is sometimes said that the Teign Valley line stations were closed as a result of The Beeching Report – but this is inaccurate. Dr Richard Beeching's famous report was published in 1963, although many of his suggestions regarding station closure did not take effect until 1965 and afterwards.

The Teign Valley may no longer exist, but to take a train ride from St David's station in Exeter, to one of the villages – Christow, Ashton, Trusham, or Dunsford for example – was a most delightful experience, and still remembered by many today. The line wound its meandering way through meadows with grazing cattle, woods, villages, hamlets, and all aspects of the Devon countryside. In those days, the smartly dressed Station Masters took a pride in their stations, and usually kept them spotless, often having small, well-tended gardens alongside.

Why, one might ask, were the days of steam so much more fascinating to most people than today's diesel?? An interesting question with probably a multitude of answers, for even today a trip on a steam train is an experience never forgotten.

The site of Alphington Halt.

The Halt finally closed in June 1958, and was later demolished.
The bridge spanning Church Road, Alphington has also disappeared.

THE EXETER TRAMWAYS

In a book on Exeter's power station, it is fitting to pay a short tribute to the tramway systems operating in Exeter for over fifty years, for it was from that building that the electric trams obtained their power.

Within a few years of City of Exeter Electric Company taking over the supply of Exeter's electricity, it was decided to improve the running of Exeter's tramways. From 1881, following the passing of the 1870 Tramways Act, Exeter had been faithfully served by horse-drawn tramcars, but with the gradual introduction of electricity to the whole country, many towns and cities were replacing horse-drawn trams with what was considered then as revolutionary electric tramcars. As early as 1901 a new company had been floated, called The Exeter and District Tramways Syndicate. It proposed, amongst other things, six miles of electrified track for Exeter, incorporating much of the existing system. That company was not able to proceed with these proposals, however, and went into liquidation.

The Exeter and District Tramways Syndicate's somewhat unexpected arrival on the scene, albeit short-lived, prompted the city council to take a serious look at the possibility of converting the existing horse-drawn tram operation to electricity. The Exeter Tramway Company was formed in 1881, and opened with three lines (a planned fourth never materialised), and just three tramcars.

Following the passing of the Exeter Corporation Tramways Act of 1903, the electrification of all the existing horse-drawn tramways in the city was authorised. In addition, there were several extensions proposed, including some that never happened – two such routes being Holloway Street and Topsham Road to Weirfield Road, and along Denmark Road to Bedford Circus via Barnfield.

On 4th April, 1905, in a High Street crowded with inquisitive onlookers, the then Mayor of Exeter, Councillor Edwin Perry, made a speech from the top of electric tramcar Number 1, outside Exeter's historic Guildhall. When that car left the High Street, Mayor Perry was given the privilege of driving it – Exeter's first electric tram. Along Heavitree Road it respectfully followed a suitably decorated horse-drawn tram, which then took its final journey to the New North Road depot and stables. The horse, so long a vital part of transport in this country, was being superseded by electricity, and of course before too long the horse would be a very rare sight in the streets of every town and city. A contemporary cartoon quipped that the horse trams had been killed by an electric shock!

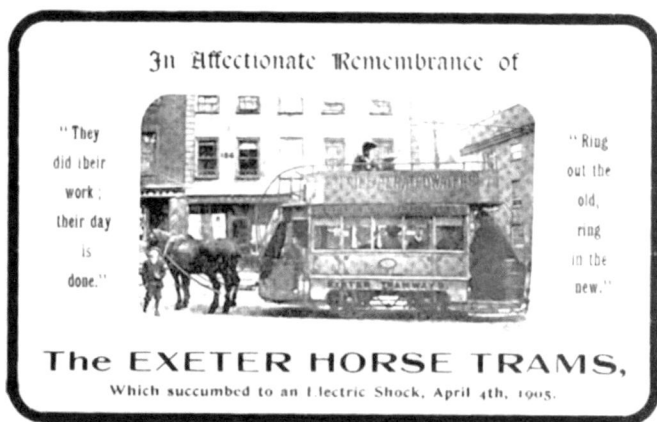

In Affectionate Remembrance of

"They did their work; their day is done."

"Ring out the old, ring in the new."

The EXETER HORSE TRAMS,
Which succumbed to an Electric Shock, April 4th, 1905.

A suitable tribute to Exeter's horse-drawn tramcars.

Over the ensuing years, Exeter's tramway system would be increased, eventually taking in routes to Cross Park Terrace (Heavitree), Pinhoe Road (Cemetery Avenue), Alphington Road and Cowick Street (St Thomas) and, more importantly, to St David's – creating a vital link with the railway station.

The New North Road tram depot was replaced with a new building in Heavitree Road (that would, eventually, become the depot for the city's motorised buses), although even that building, located next to the Exeter Swimming Baths, has now been replaced by housing.

Such was the impact of the electric tram that even Exeter's Exe Bridge was demolished and a new, wider bridge constructed to accommodate the city's trams. The previous bridge, opened in 1778, had been designed with three arches, each arch supported by huge piers buried deep in the riverbed. However, the piers proved an obstruction for the river, and in severe weather it hampered the flow to such extent that flooding was reasonably commonplace in the St Thomas area.

For over one hundred years there had been discussions within the city council regarding the possibilities of a new Exe Bridge – but the old bridge had remained whilst the talking continued, and it was not until 1900 that councillors finally agreed to construct a new bridge spanning the river. The more recent discussions regarding a new bridge coincided, of course, with other discussions in the council chambers regarding the possibility of Exeter having electric trams. The existing bridge would not be suitable, in that it was too "arched", and also not wide enough to take two lines of tracks, necessary for the electric system.

The contract for the design of the new Exe Bridge was awarded to Sir John Wolfe Barry KCB, (who designed of Tower Bridge in London) and Cuthbert Bretherton MinstCE. It was some ten feet wider, and considerably flatter than its predecessor, and had just a single arch instead of three. The roadway across the bridge was constructed in concrete, with an asphalt layer on to which was placed wood block pavoirs surrounding the new tramlines. Footpaths either side were almost eight feet wide, much wider than those of the first bridge.

In July 1904, the Mayor of Exeter, Councillor F.J. Widgery JP, laid the corner-stone, and in March 1905, the new Exe Bridge was officially opened by the Mayor for that year, Councillor Edwin Perry.

Exe Bridge pre-1905.

The 1905 replacement bridge, designed wider and shallower than the previous bridge.

At various locations around the city, even as late as the 1960s, there were small green metal boxes, known as breaker boxes (although sometimes referred to as section pillars), each being embossed with the Exeter coat of arms. The breaker boxes were set at half-mile intervals, with the underground cables coming up to a junction in each box, before continuing to the next box.

Thus, along the whole of the route the cables took, any area could be switched off within half a mile of a problem. This meant that the whole of the system did not need to be cut off at the same time, and areas could be isolated – quite an ingenious arrangement.

One of the city's several breaker boxes (or section pillars).

It is thought that there were probably less than a dozen such boxes in Exeter.

The breaker boxes also acted as distribution boxes when electric cables were required to go in another direction. The cables supplying power to the trams were obviously separate to those supplying normal power, but also came into the breaker boxes, and it follows that given areas of the overhead tram cables could also be isolated. In addition, all of the section pillars were connected by telephone to both the generating station and the tram depot in Heavitree Road (that building having been demolished many years since).

One typical breaker box, shown in the photograph above, was located in Gervase Avenue. It was photographed as it was about to be removed, for display at the Royal Albert Memorial Museum. A former employee of that Museum, a gentleman by name of Harris, was a keen tram enthusiast. Realising the significance of the breaker boxes, he fortunately persuaded the authorities to preserve this particular box. It is thought that it could be the only one from Exeter now surviving.

The tram power cables were all double-insulated, and distributed around the city in underground conduits. At intervals, the cables were taken up steel poles to arms that extended over the streets, whereby trams were connected to the supply throughout the various routes. The poles were usually known simply as tram poles, or sometimes as traction standards, and were erected in such a way to obtain the most efficient support for the overhead cables. There was no set distance apart, and on curves in the road, or at junctions, more would be required than on a straight section.

To counteract the sagging of the overhead cables, poles were never set more than 120 feet apart, and more often than not, much closer. In some instances, the cables were also supported from adjacent buildings, where the conductor wire was suspended from another wire, which itself was attached to "roses" on buildings either side of the street. These "roses" were quite ornate, but none is thought to have survived.

The Exeter electric trams had proved most successful for twenty-six years, yet times change, of course, and progress is made. Even before 1930 the city council had been looking at the possibility of changing from electric trams to motor-driven vehicles. The end was in sight for the trams, although they, as with their four-legged predecessors, had served the city well.

Seven new motor buses were delivered to Exeter in March 1929, two years before the trams ceased. They were all single-deck – two Leylands, two Bristols and three Maudslays. The following year, a further six Maudslays arrived, together with a Commer. The Commer proved a disaster and was later replaced with another Maudslay.

January 1931 saw the arrival of six AEC Regent double deck buses, followed by another nine Regents, and ten Leyland Titan double-decks in August.

Exeter Electric Tram No 10.

Photographed in Alphington Road. Tram poles, with decorative supporting brackets, carry the overhead cables to which trams were connected.

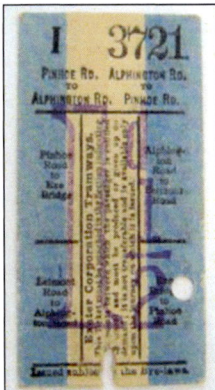

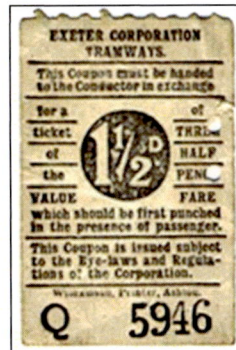

Tickets from the Exeter Corporation Tramways.

The centre ticket shows a typical advertisement printed on the reverse.
Cornish & Co. was, for many years, one of Exeter's most popular outfitters.

In August 1931, the end came for Exeter's trams, and once again Councillor Perry, though no longer Mayor, had the honour of driving the last tram from the city centre to the Heavitree Road depot. This time, however, he was driving the leading vehicle and was followed by one of the new additions to Exeter's transport fleet – a Regent double-deck omnibus. When the last horse-drawn tram left High Street, it turned left into New North Road. On this occasion Councillor Perry directed his tram down Paris Street and into Heavitree Road, where it entered the depot for the last time. The Exeter Corporation Tramways had reached its final destination, and yet another era in Exeter's transport system had passed into oblivion. The age of the motor vehicle had finally arrived, and since then, no tramcars have been seen in the streets of Exeter.

Some Exeter trams were sold to Halifax and Plymouth tramway companies, but most fell into the hands of local scrap merchants and were broken up. For many years the bodies of several could be seen in private gardens, used as very convenient garden sheds, or for storage, and indeed there may well be some remnants of the former tramcars still in existence to this day.

The only remaining tramcar of the Exeter fleet is Car Number 19, a vehicle constructed in 1906, and used on the Exeter tramway system until the closure in 1931. It was then abandoned, but eventually recovered and taken on by the Seaton Tramways. It has been restored and is still in its original width form, which makes it look slightly unwieldy on the narrow gauge system at Seaton. It has no upper deck today, as that would have made it too tall for the Seaton depot – and far too unstable to carry passengers upstairs. It is no longer in constant use, being brought out on special occasions as a visitor attraction at the Devon town.

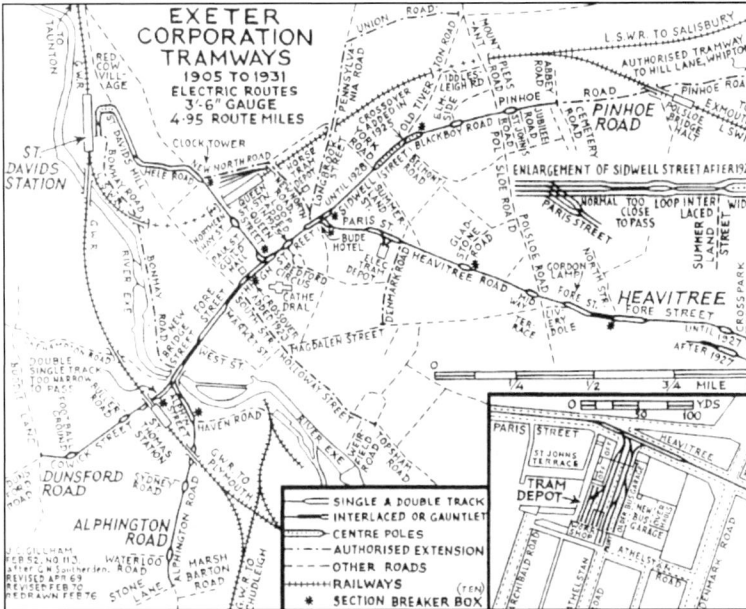

The layout of the Exeter Corporation Tramways.

Sectional view of a tram line.

This has been cut and sand-blasted.
Originally it was part of the track
in Queen Street.

Exeter's last tram entering the Heavitree Road depot in 1931.

The buildings in the background still exist, but the Heavitree Road Depot was demolished during the 1960s, and a new Bus and Coach Station opened in nearby Paris Street.

THE END OF THE ROAD

Electricity continued to be supplied from Haven Road for many years until 1947,when virtually all of the independent and council-operated electricity companies were nationalised. In the southwest of England, the South West Electricity Board came into being – in common with several other such regional boards throughout the country. There followed the inevitable closure of former electricity generating stations, and many were either demolished, or put to other uses. It may be of interest to note that in 1946 the Exeter Corporation Electricity Undertaking sold 69,499,470 units of electricity. The following year it sold 78,005,492 – an increase of over eight million units. That one year's difference shows how popular electricity had become, and with electricity approximately three-quarters of one penny per unit in 1947, that represents something in the region of an increase of £25,500 in the company's profits. That was quite a substantial sum in those days.

The building of the National Grid, in 1929/1930, enabled the task of shutting down the smaller, inefficient power stations to begin. New, much larger stations were being constructed and these could feed the grid far more efficiently. This, in turn, enabled the price of electricity to be brought down, so that the high-voltage circuits involving greater costs could be extended by all undertakings into rural areas. (This process went on well after Nationalisation, culminating in a successful rural development programme.)

Alternative methods of supplying electricity have replaced most of the old power stations. The nuclear powered stations at Hinckley Point, in Somerset, now provide electricity to the southwest of England, and of course experiments continue with wind farms and other means of creating electricity. Hinckley Point "A" started supplying electricity in 1965, and continued for thirty-five years. Hinckley Point "B" started operation in 1976, and although it was due for decommissioning in 2011, in 2007 British Energy was given a further ten years to operate it.

The Exeter Electric Light & Power Station in Haven Road was closed by the Central Electricity Generating Board on 23rd March 1960. For the previous twelve months, its main function had been an emergency source of power during the transfer to the new National Grid supply operated in this area by the new South Western Electricity Board (SWEB), but the construction of the grid system, supplied by more efficient power stations, ensured the replacement of all the old power stations. Staff from Haven Road had either moved over to this new company, or been transferred to other power stations. A few opted to take jobs in other fields, away from electricity, although others were temporarily retained at Haven Road to keep maintain the building should it be required again.

Some of the staff had been employed at the power station for most of their working lives, and in one or two instances employees had never been employed elsewhere.

In September 1961, the Editor of *The Power News*, then a journal of the Central Electricity Generating Board, received a letter in response to an article concerning a man who had been employed at a power station for forty years.

The letter was from Percy Marks, who lived at Okehampton Road in Exeter. In the letter, Percy explained how he had started at the power station in Exeter in 1945 and remained there until he retired in July of 1961 – a period of service spanning over forty-five years. He was the longest serving employee at the Exeter works. He recalled the early days when reciprocating engines were being used prior to the five turbines being installed during the 1920s, and then remembers the turbines being demolished as scrap in 1960, along with the boiler house plant, and the coal plant, after the station's closure.

Percy explained in his letter that the main chimney was demolished brick by brick, not blown up, and from May 1960, when the station superintendent was transferred to another power station, Percy became in sole charge of the demolition work. Percy Marks was one of the employees who were on duty during Exeter's main air raid in 1942, and helped to maintain the supply of power to the City under difficult conditions – including a fire within the premises.

Other long-serving employees included Fred Anning who had served alongside Percy Marks during the Exeter blitz. (The Electricity Undertaking had also employed Fred's father as a Mains Superintendent.) Bill and George Phillips were employed on the coal conveyor plant, and they were continuing an association with the power station that had started with their grandfather in 1914.

Three of the turbine attendants could also boast of many years service between them. Jack Sculpher had worked at Haven Road for thirty-nine years when it closed; Harry Barnes had done so for thirty-two years, and Fred Kelly for thirty years.

Following the closure of the power station, the huge chimney was taken down by contractors working for local demolition expert, Alf Carpenter. In March, 2007, the *Express & Echo* reported that the copper tip of the lightning conductor was given to Tony Johns of Ide Lane, Alphington. Tony's father was at that time working for Carpenter's company. The remaining section of the conductor, some 150 feet in length, was, of course, valuable scrap metal for the company. Tony Johns said in the article that, as a young man in his late teens, he watched a demolition expert climb to the top of the chimney, and start swinging a sledgehammer, knocking the brickwork down brick by brick, into the chimney itself. Tony told the reporter: "you would not get away with that today, Health and Safety would be all over you". He was, of course, quite right!

All of the machinery was removed from the interior, much of the work being done by employees still working at the power station, and a short time after, a large part of the building was demolished.

From 1961, the former offices and engine house were all that remained of the original building. For several years it was retained by the South Western Electricity Board and used as a store, mainly for heavy switchgear and cable reels. At some stage during the 1970s, the buildings behind the northwest end of the twentieth century extension were replaced by an electricity sub-station, which is still in existence today. Despite the renovation work currently taking place (in 2008), it is assumed that this will be required to remain in place.

The offices were occupied by store managers, store keepers and store administration staff. Based elsewhere in the building were electricians, underground workforce and voltage pressure staff. In the early 1970s contracting repair administration staff were also based there, using what was known as "The Wheel System" for holding small job order repair documents. All of these employees and equipment were later moved to the SWEB depot at Sowton, on the outskirts of Exeter, and as the Board had no further use for the building, it stood empty and unwanted. Eventually it was taken over as part of the Exeter Maritime Museum, although when that part of Exeter's tourism industry was lost in 1997, the power station became empty once again.

Inevitably, being an empty building, it attracted those who, for some reason, enjoy wrecking other people's property and creating as much havoc as they are capable of doing. Many of the windows were broken, some – as has already been said – containing the original stained glass. It also attracted those with an eye to make quick money, and the old Victorian-style fire grates and mantel surrounds were stolen. Sadly, for several years, nothing was done to stop any of this happening, and the building was almost totally ignored.

The *Express & Echo* report
of the Power Station
closure in March 1960.

Photograph courtesy
The Express & Echo

The turbines being removed from the Engine House in 1961.
Photograph courtesy SWEHS

The last of the boilers awaiting removal to the scrap yard.

Demolition of the two chimneys.
Left, a birds-eye view of the base of the brick chimney under a pneumatic drill.
On the right, the last section of the metal chimney is about to go.
Photographs courtesy SWEHS

With the continual development and enhancement in the Exeter Quay and Basin vicinity, there was, at one time, a possibility that the building would be demolished, probably to make way for the inevitable blocks of flats or shopping malls, that have taken over much of the city in recent years. It was not until the city council fortunately decided not to allow demolition, preferring instead to seek ideas for the revival of Exeter's former power station, that all of the windows were covered with protective boarding – albeit too late to save some of the attractive stained glass windows of the offices. Fortunately, the interior did not suffer too much damage by vandalism, although dry and wet rot have taken their toll in some of the offices and stores.

Apart from thoughts of demolition and housing development, many other suggestions were made for the building, including one by transport enthusiast Colin Shears, who thought it was ideal for a possible transport museum – which, although ignored at the time, seems to have been a somewhat appropriate and sensible suggestion for a building that had supplied power to the Exeter trams for over thirty years. Even that idea never came to fruition, and it remained an empty shell.

One of the empty first-floor offices.
Decay, vandalism and neglect make this a sad sight.

THE FUTURE

After years of neglect, the city council eventually invited proposals to be put forward for the conversion of the building. One scheme included designs from the Millhouse Partnership, and it was those proposals that were finally approved by the council planning committee, and accepted in preference to any others.

The Millhouse designs included using the former engine house as an art gallery and function room, with a small, eight-bedroom hotel and a small restaurant in the area formerly used as offices. On the land to the south of the building, Millhouse proposed the construction of seven self-contained apartments designed, it was said, to complement the old building. It is frequently a matter for discussion when modern design should be linked to early architecture, and it may be that some traditionalists will say that the flats should have followed the design of the original building more closely. Others, perhaps, will be pleased at the juxtaposition of two completely different designs. It has to be left to the eye of the beholder.

Following a delay caused by the need to remove asbestos from the building, construction work eventually began in January of 2007, and within a short time the building once again came to life, buzzing with creative activity in preparation for its new life.

Project Manager Trevor Herbert recalls that during the groundwork excavations for the new apartments, the developers encountered some difficulty, in that the existing underground services for gas, mains water, electricity and surface water for the modern buildings in the surrounding area, were all in positions where steel piles were required to be driven down, in order to stabilise the new apartments alongside the power station.

Every one of these services required re-routing, adding to the time and cost. Having diverted the services, the piles were then able to be driven into the ground, to a depth of six metres in most cases, and filled with concrete. The foundations were built around them before a concrete slab was placed in position to carry the weight of the new building. It has to be remembered that a great deal of the foundations would be at, or possibly below, the level of the Basin and River Exe, and thus stability was a vital ingredient of the groundwork design.

During the excavations, a nine-inch pipe was found running alongside, and parallel to, the former power station. It is thought that this could have been one of the original pipes recycling the excess water from the boiler house back to the Basin. Unfortunately it was removed before it was realised that an original pipe could have been found in that vicinity.

The architect's drawing for the apartments.

Photograph courtesy Millhouse Partnership

On exterior photographs of the power station (such as that shown on the front cover), can be seen the large power station chimney, and, once again, excavations for the apartments uncovered another fragment of the old building – the base of the original chimney structure. Even some of the elaborate, angled brickwork was still in position on the section of chimney immediately above the concrete base.

The vast concrete slab on which the chimney stood was set deep enough for the contractors to allow it to remain *in situ,* without obstructing the new foundations. This was perhaps fortunate, for the whole slab would originally have been considerably larger than the chimney base, and designed to withstand the enormous weight and pressure of a chimney 150 feet high, and would have been difficult, and costly, to remove.

Following the completion of the apartments, work has started on the conversion of the former power station building. However, the plans for restoring the old building were somewhat hampered in March of 2007, when arsonists entered the building and started several small fires around equipment stored in the former engine house.

Some of the fires caused very little damage, but others set fire to insulation material, which when burning gives off a dense, acrid, smoke. Apart from destroying several thousand pounds worth of materials, the fires caused considerable smoke damage to the ground floor. In addition, the Project Manager's office was completely trashed, with valuable plans, equipment and personal items being destroyed.

Totally senseless vandalism, known to be the work of youngsters, caused unnecessary expense and delay in the project, although fortunately no serious damage was caused to the structure.

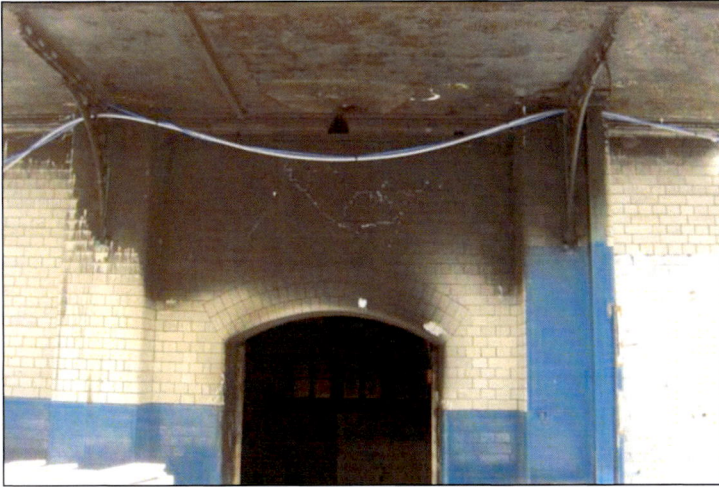

Damage caused by arson in 2007.
The only damage to the structure was caused by smoke, but thousands
of pounds worth of materials and equipment were lost

The new apartments alongside the old building, 2008.

Thus, after standing on the site for over one hundred years, the Exeter Electric Light & Power Station now begins a new lease of life, and it has been designed to serve a useful purpose once again, although towards the end of 2008 the country's economic crisis caused work on the building to be temporarily halted. Despite that, Millhouse Partnership has thankfully assured the building's future, and the building will doubtless be enjoyed by both residents and tourists for many decades ahead.

In a historic city such as Exeter, it is satisfying to know that at least *some* buildings are being re-generated, rather then simply demolished and wiped out of history – as, regrettably, has been the fate of many other buildings in the city over the years.

The Exeter Electric Light & Power Station.
Seen from Colleton Crescent in July 2008. The new apartments are complete and the old building awaits renovation.

The Basin area, 1945

This photograph, taken by the R.A.F, shows the square-ended City Basin, with the River Exe at the bottom. The Basin Branch Line railway track runs from the top to bottom of the photograph. The power station is the three-gabled building just below centre, with large heaps of waste coke visible above it. The two chimneys are visible either side of the building.

Photograph courtesy National Monuments Record (English Heritage)

APPENDICES

APPENDIX A

ELECTRICITY HOUSE, FORE STREET, EXETER

APPENDIX B

EXTRACTS FROM MINUTES OF THE ELECTRICITY
COMMITTEE, EXETER CITY COUNCIL, 1937 – 1948

APPENDIX C

EXTRACTS FROM RANDOM ISSUES OF THE JOURNAL
THE ELECTRICIAN, 1882 –1904

APPENDIX A

ELECTRICITY HOUSE, FORE STREET, EXETER

The main entrance of Electricity House in Fore Street.

With the closing down of the Rockfield Works in 1904, it was deemed necessary to move the administration side of the City of Exeter Electricity Undertaking to new premises in nearby Sidwell Street.

In 1928, to cope with the development of electric cooking, water heating, and the ever-growing other domestic and commercial applications of electricity, additional premises, adjacent to the offices, were secured and adapted as showrooms and stores. By 1930, the administrative and distribution technical staff had outgrown the office premises in Sidwell Street, and some of the departments moved into larger, more suitable, premises in Dix's Field.

By 1934, it was found that even the combined accommodation available in both Dix's Field and Sidwell Street was becoming quite inadequate to house the staff necessary to deal with the enormous increase in the volume of business. Consequently an existing building of suitable size in Fore

Street, at the heart of the city, was purchased, and arrangements were made to carry out extensive structural alterations, etc., to adapt it for the needs of the Electricity Undertaking.

It was, quite appropriately, named *Electricity House,* and it seems that expense was not spared in the design, with much bronze and marble being used, together with new systems of heating and lighting. The detailed plans also show that a considerable amount of walnut panelling was used in the showroom and other areas, particularly those that had public access.

On the afternoon of Wednesday, 7th October 1936, the new offices and showrooms were opened with considerable pride by Sir John Brooke, CB, Vice-Chairman of the Electricity Commissioners. Also present were the Mayor of Exeter, Alderman Percy Gayton and Mrs Gayton, and the Chairman of the City Council's Electricity Committee, Councillor Arthur Brock MC., plus numerous other dignitaries.

It was anticipated that the new premises in Fore Street would provide the ideal answer. The Electricity Undertaking's offices, stores, showrooms and demonstration theatres would now be under one roof.

Yet the Undertaking's problems were far from over. Just six years later, during the blitz on Exeter in the Second World War, large tracts of the central part of the city were devastated. Buildings were destroyed, either by direct hits from enemy high-explosive bombs, or the resulting fires caused by such action, with many others also lost in fires caused by the smaller incendiary bombs. Others remained standing, but the dangerous state of many ensured they would never be used again. Much of the city centre, especially South Street and the top end of Fore Street, was reduced to rubble.

One of many buildings totally destroyed was Electricity House. So ended another unfortunate phase in the company's administrative life.

The following pages contain extracts from the official souvenir brochure issued at the opening ceremony of Electricity House.

THE CITY OF EXETER ELECTRICITY UNDERTAKING

HISTORICAL NOTES

The Public Supply of Electricity in Exeter was originally undertaken by a local company who commenced operations in 1889. The Company consisted primarily of residents of the City, and much credit is due to those pioneers who financed the Undertaking in its early days.

Seven years after the commencement of the supply, it was found that heavy additional capital would be required to enable the business to be developed on satisfactory lines. The Company, however, found it impossible to raise this capital, and consequently the concern was disposed of in 1896. It must have been a matter of considerable satisfaction to the local shareholders that their own City Council became the purchasers. Since that date, the business built up on the small but satisfactory foundations laid by the Company has developed into a magnificent example of civic enterprise.

The citizens of Exeter own today an Electricity Undertaking of which to be proud. It consists of a modern Generating Station and nearly 300 miles of overhead and underground transmission and distribution cables. These cables form a large supply network covering an area of 41 square miles, which is served by 150 transforming sub-stations, distributing to over 19,000 consumers. On account of the high efficiency and low costs of production of the Power Station it has been selected by the Central Electricity Board for operation in conjunction with the National Grid Scheme. The supply network supplies the City of Exeter and eight surrounding parishes, and in addition, provides a bulk supply of electricity to the Exe Valley and East Devon Supply Companies, who serve an area of approximately 620 square miles, north, south and east of the City.

The original Generating Station was situated in New North Road, adjacent to the Bridge over the Southern Railway, and was equipped with watertube boilers, horizontal engines (non-condensing) rope-coupled to single phase Morley alternators, and vertical high speed engines belt-coupled to

Thomson-Houston arc lighting dynamos. The total capacity of this early plant amounted to only 220kWs.

After the acquisition of the Undertaking by the City Council, various additions to the plant between the years 1896 and 1899 brought the capacity up to 675kWs. The increased capacity did not prove adequate for very long, and in 1904 the present Power Station was put into commission on a site at the City basin. Here there was ample room for future expansion (3¾ acres), a plentiful supply of water for condensing purposes and railway facilities for the delivery of coal directly into the Power Station,

The new Station, built and equipped on the most approved principles of the day, had a capacity of 1,500kWs, but the buildings were designed with a view to the ultimate accommodation of much larger units of plant.

By the year 1922, the capacity of the Station had been increased to about 4,000kWs, and between the years 1925 and 1930 further extensions followed in quick succession, bringing the total capacity up to 15,350kWs.

The more recent of these extensions was designed for three-phase operation at 6,600 volts and the standard frequency of 50 cycles per second, the general change-over of the system of supply from 60 cycles to 50 cycles having been carried out in 1928.

To make room for these modern additions, the whole of the generating plant which had been installed before 1915 was removed together with the original boilers and switchgear.

The Generating Station is now a model of modern practice and operating efficiency, and for these reasons was, as previously mentioned, made a Selected Station by the Central Electricity Board under the provisions of the 1926 Act.

SOME COMPARIONS TO SHOW THE GROWTH
OF THE UNDERTAKING

NUMBER OF CONSUMERS:
Year ending 31st March 1900 350

Year ending 31st March 1936 18,463

SALES OF ELECTRICITY:
Year ending 31st March 1900 .... 278,000 units

Year ending 31st March 1936 .... 24,715,625 units

TOTAL REVENUE:
Year ending 31st March 1900 .... £   6,424

Year ending 31st March 1936 .... £163,020

AVERAGE PRICE OBTAINED PER UNIT SOLD:
Year ending 31st March 1900 .... 5.55 pence

Year ending 31st March 1936 .... 1.45 pence

STAFF & WORKMEN EMPLOYED:
Year ending 31st March 1900 15

Year ending 31st March 1936 149

CAPITAL EXPENDITURE:
Year ending 31st March 1900 .... £   25,572

Year ending 31st March 1936 .... £751,962

EXAMPLES OF ELECTRICITY COSTS FOR 1936

Item	Price per unit
Domestic lighting (Flat rate)	½p
Business lighting, sliding scale according to quantity	4d to 2¼d
After hours shop window lighting & illuminated advertisements	2d
Heating	1½d
Heating (controlled)	½d
Cooking, or cooking & heating combined	1d
Water Heating	1d
Water Heating (controlled)	½d
Power, sliding scale according to quantity	1¼d to ¾d
Power (controlled)	¾d

Electricity House - The Demonstration Theatre.

TWO-PART TARIFF SYSTEM

For Private Residences (for all domestic purposes):

1. A unit charge of ¾d per unit for the first 100 units per quarter, and ½d per unit for all units in excess of the first.

2. *Plus* a Fixed Charge based on net Rateable Value of the house:
   Up to £12RV: 30/- per annum
   From £12 to £30   add 2/- in the pound
   From £30 to £40,   add 1/6d in the pound
   From £40 upwards,   add 1/- in the pound

For Business Premises:

1. A unit charge of ½per unit, plus

2. A Kilowatt charge of £3 per quarter per kilowatt of maximum demand

Electricity House - The Showroom.

THE FORE STREET OFFICES AND SHOWROOMS

The premises are "T" shaped, with the main frontage to Fore Street, a side frontage to South Street, and an entrance from George Street.

The buildings were originally a tobacco factory, and consist of a basement, ground floor, first floor, second floor and third floor, having a total area of approximately 20,300 square feet.

The buildings, when taken over by the Council, were of sound construction, but needed entire remodelling, including the alteration and removal of a good deal of the steel and cast-iron work of which the floors were mainly constructed.

The following accommodation has now been provided:

Basement Floor:

This floor is used for stores, and is divided into various compartments by open metalwork screens. There is space for packing and unpacking, with a goods lift serving the basement, ground, first and second floors. A strong room and sub-station have also been provided.

Ground Floor:

The Fore Street frontage, for a depth of 20 feet, is devoted to the Main Entrance and display window. Immediately behind the display window the showroom occupies the whole width of the premises (29 feet) and has a depth of 44 feet. On one side of the showroom is the Accounts department, and on the other a counter for "Enquiries", a model kitchen and a model bathroom, and between the latter two an automatic passenger lift operates to the first, second and third floors. The Demonstration Theatre is entered from the showroom and has a platform seating for 64 people. Behind the showroom are various offices etc.

First Floor:

On this floor, two waiting rooms are provided for general use. There are also rooms for the Mains Superintendent, Overhead Lines Superintendent, technical assistants, meter store, sundry stores, Mains Engineers, showroom workshop, meter fixers, meter testing, staff lavatories etc.

Second Floor:

Front portion: Living room, sitting room, kitchenette and offices for Mains Foreman.

Rear portion: Drawing office, Printing room, Cooker repairs department, Water heater testing and repairs, Records & stationery and sundry stores.

Third Floor:

Front portion: 3 Bedrooms, bathroom, etc. for Mains Foreman.
Rear portion: City Electrical Engineer's office, Waiting room, Chief clerk, General clerks, Typists, Ledger clerks and Mailing office.

Display Windows & Arcade:

The main entrance is in the centre of the front. This is enclosed with a bold surround in bronze to the underside of the existing shop cornice. In this surround *"Electricity House"* is formed in 12inch letters, placed in an illuminated coloured glass fascia. The surround has illuminated bosses and insets.

The windows on each side of the Main Entrance are of plate glass, 7ft.0on wide and 7ft.6in. high. Above these are embossed and illuminated glass panels embodying the City Arms.

The Arcade from the street line to the main entrance doors has plate glass and bronze framing on both sides.
The base of the Display Windows and Arcade is in emerald green granite.
The vestibule doors are in walnut, with a surround of Tinos marble
The Arcade is floored in Travertine marble, and illuminated by concealed cornice lighting.
For the two display windows, a dual system of lighting is in use, *i.e.* white and coloured.

Showroom:

The walls are lined up to a height of 9 feet in walnut panelling of bold design. This panelling is carried around the outside of the model kitchen, and the model bathroom, and the counters for the accounts department and "Enquiries" are treated in a similar manner.
The interiors of the model kitchen and the model bathroom are lined with Vitrolite, the colours being primrose and green.

Heating:

Generally, the offices are electrically heated on the tubular system, with provision at certain points for radiators. The Showroom is heated by specially designed concealed electric heaters. The systems as a whole are controlled thermostatically, and in addition governed by a time switch.

APPENDIX B

EXTRACTS FROM MINUTES OF THE
ELECTRICITY COMMITTEE, EXETER CITY COUNCIL

THESE RANDOM EXTRACTS RELATE TO THE HAVEN ROAD POWER STATION, AND ARE FOR A TEN-YEAR PERIOD, COVERING THE SECOND WORLD WAR YEARS.

January 1937 Resolved to sell 1.86 acres of land at the rear of the Power Station to Mesrs Garton & King of Water Lane.

January 1938 Agreed to accept 10,000 tons of Tredegar Oakdale coal from William Cory of Wales, at thirty shillings and eight pence per ton. Approximately 750 tons of coal per week were used at the Power Station at that time.

January 1938 Applications were invited for a Switchboard Attendant at £4.4s.0d per week, as per the current National Scale.

February 1939 Following the resignation of Mr Roscoe, the Switchboard Attendant's position was again advertised, at an increased wage of £4.7s.0d per week.

January 1943 Agreed to increase the wage of Mr Hardman, foreman fitter by 4d per hour, taking his wage to £5.0s.9d per week.

April 1943 The committee had approved a contract with Messrs Fraser & Chalmers in 1940 for the supply of new coal handling plant. It was reported that the installation had now been completed at a cost of £10,934.

December 1943 Owing to the serious deterioration of the slate roof of the Engine House, laid in 1904, water had been entering the building. Repair work had been carried out, but as identical slates could not be found, asbestos sheets had been used, but suitably camouflaged.

December 1944 The Ministry of Power had suspended supplies of Welsh coal as it was required for important war industries in occupied countries. The replacement was Birch Coppice coal from Warwickshire, which although slightly cheaper per ton, had less heat value and therefore needed more coal to maintain the same heat output. The estimated additional cost for 1945 would be £30,000 as one third more coal would be required.

March 1945 The committee were informed that an article in The Express & Echo recently, it was stated that the number of consumers of gas in the city had increased by 36% in recent years. The committee was told that in the same period, the number of electricity users had increased from 3,868 to 17,100 – an increase of 340%.

September 1945 It was reported that the blast walls erected at the Power Station to protect the machinery had now been removed.

January 1948 The committee was informed that the new switchgear for the switch gallery, purchased from Messrs Cooke & Ferguson, will be completed at the end of January 1948.

March 1948 The committee agreed to purchase a Nissen Hut for storage, to be located alongside the Power Station on the Exe Bridge side.

Nissen huts on land adjacent to the power station.

Although the date of these photographs is not known, one of these huts is possibly that referred to in the Council minutes above.

The top photograph shows the offices, engine house and boiler house, but the coal bunker house appears to have been demolished. The houses in the bottom photograph are in Diamond Road.

APPENDIX C

EXTRACTS FROM THE TRADE JOURNAL "THE ELECTRICIAN"
Taken from the collection at Bristol University, by P.G. Lamb, secretary, S.W.E.H.S

21.1.1882   A proposal from a private company to light Exeter with electric light has been referred to the Streets Committee of the Council. It is proposed to utilise water power, of which there is an abundance in the neighbourhood.

11.9.1885   Tender accepted for electric light to be installed at the new Asylum under construction.
(Note: This must refer to the Digby Hospital, opened on the outskirts of Exeter in 1886)

18.2.1887   Theatre at Exeter is to receive an installation of arc lights. *(Note: It never did receive them. The Theatre Royal in Exeter was gutted by fire in September, 1887)*

18.2.1887   Courtlands House, at Lympstone, near Exmouth, the residence of Mr William Lethbridge, was supplied with a plant of about 100 lights by Messrs Crompton & Co Ltd. *(This is thought to have been the first house in Devon to have electricity)*

2.12.1887   An informal and private meeting, convened by circular, was held at The Victoria Hall to discuss the propriety of forming a company for the purpose of supplying Exeter with electric light.
Meanwhile, Mr Massingham proposes with the consent of the authorities, the erection of wires to light experimentally a central area of Exeter, the Albert Museum, the Athenaeum and the establishments of several tradesmen.

23.12.1887   Preparations for the introduction of the Thomson-Houston system are being actively pursued in Exeter. The fixing of the poles and wires in High Street, Queen Street and a portion of St Sidwell's is nearly complete.

3.2.1888   The citizens of Exeter do not however, as a whole, take kindly [to the arc lights], the principal objection being that the light is too glaring for the eyes. This is possibly the fault of the globe.

10.2.1888   The lighting of the streets in Exeter by electricity is a decided success. The arc lamps are hung at a much greater height on poles than was the case when a few trading establishments were being illuminated some weeks ago.

6.4.1888   At the meeting of the Exeter Town Council, a letter was read from Mr W.H. Michelmore, secretary of the Exeter Electric Light Company, notifying the intention of the directors to submit a tender for lighting the principal streets of Exeter by electricity.

27.4.1888   Mr Massingham, as agent for the Thomson-Houston system, attended the Council meeting proposing 40 arc lamps at £800 per year (£20 per lamp). Not adopted.

8.6.1888   The Exeter Electric Light Company have commenced operations at Trew's Weir on the River Exe, where the central station will be established
(Note: this refers to the central generating station, not the Central Station of the railway)

20.7.1888   A meeting of the Committee of Exeter Town Council was informed that notice had been given both to the owners of Trew's Weir and to the Electric Light Company.
The Exeter Electric Light Company had purchased The Rockfield Hat block of buildings in New North Road and machinery is to be erected at once for the supply of private houses in the City.

2.11.1888   At a meeting of the Exeter Town Council, the Committee opened tenders as follows: *[there then follows a list of various tenders from companies wishing to provide Exeter with equipment to supply Exeter with electric light].* Having regard to the large increase in the cost as compared with gas, the Committee did not recommend that any of the tenders be accepted.

17.5.1889   The Exeter Electric Light Company has offered to tradesmen first, and private residents next, a supply of light free of charge for two months. Much progress has been made with the work of erecting machinery at the central station in New North Road.
(Note: Again, this refers to the central generating station, not the railway's New North Road 'Central Station')

31.5.1889   The Town Council has given permission for the Exeter Electric Light Company to erect ninety-nine posts in positions approved by the City Surveyor.

4.10.1889   The motion for cancelling the permission recently given to the Exeter Electric Light Company to erect overhead wires was rejected.
The Exeter Electric Light Company was said to have successfully installed electric light at the new Exeter Theatre Royal, the installation comprising 450 Edison-Swan lamps.
(Note: This refers to the theatre rebuilt following the fire. The Theatre Directors, however, claimed 510 new electric lights when the premises re-opened)
Over 1,000 lamps had been ordered for private houses, shops, hotels and public institutions at Exeter, and the Company will complete their station next month.

15.11.1889   The Company ran their arc lamps on Monday last for the first time.

25.4.1890   The Exeter Company have decided to apply for a provisional order and place its wires underground.

9.1.1891   A meeting of the ratepayers was held on Tuesday evening to protest against the proposal recently made in the City Council to expend £30,000 on works for the supply of electricity etc.

20.5.1892   At the last meeting of the Town Council a resolution was passed stating that it was desirable to light the open spaces of the City by electric light, provided that it could be done at reasonable cost.

2.9.1892   The process of changing the system of electric light from overhead wires to that of underground mains has been proceeding for the past seven weeks and is expected to be completed in three weeks.

14.10.1892   Several non-conformist places of worship in Exeter and district have adapted the use of electric light, the most recent being the large United Methodist Free Church.

5.5.1893   The Devon Standing Joint Committee has discussed lighting its offices. It has ultimately decided to light the County Clerk's offices.

2.6.1893   The Devon Standing Joint Committee has decided to light Exeter Castle by electric light.

30.8.1893   A special meeting of the City Council had discussion on purchasing the Electric Light Company Limited, the critical condition of the Company rendered it necessary that an immediate decision was required. £10,000 was offered.

27.7.1893   The [Royal Albert Memorial] Museum new wing to be lighted electrically.

27.7.1893   The [Royal Albert Memorial] Museum new wing to be lighted electrically.

4.10.1899   City Council agreed for a seal to be fixed to purchase the Exeter Electric Light Company. It also agreed to apply to the L.G.B. for a loan of £11,000.

18.10.1899   After consultation with Dr Fleming, agreed to offer £7,500 for the Exeter Electric Light Company.

21.7.1899   The salary of Borough Electrical Engineer (Mr H.D, Munro) was increased from £200 to £275.

19.2.1904   The salary of the City Electrical Engineer (Mr H.D. Munro) has been increased to £425 per annum.

ACKNOWLEDGMENTS

When I started research into the Haven Road Power Station, one of the avenues of that research led me to **The South Western Electricity Historical Society**. At their headquarters in Bristol, the Society maintains an amazing documentary archive relating to the history of electricity in this part of the country. It also displays a fascinating museum of electrical artefacts of all descriptions, which can be viewed by the public. The support of this Society in the preparation of this book is acknowledged, and in particular the help and constructive criticism offered so readily by the S.W.E.H.S secretary, Peter Lamb.

In the hope that I have not omitted anyone, I also wish to acknowledge the assistance given by the following (in no particular order) in the preparation of this publication. To those who have been inadvertently left out, my sincere apologies

Robert Crawley *Director & Hon. Archivist, The West Country Historical Omnibus & Transport Trust*
www.busmuseum.org.uk

David Cornforth *Exeter Memories website*
www.exetermemories.co.uk

Geoffrey Harding *The Exeter Local History Society*
www.exeterhistorysociety.co.uk

Trevor Herbert *Haven Road Project Manager for Millhouse Partnership*

Paul Israel *Thomas A. Edison Papers, State University of New Jersey, USA*

Millhouse Partnership

Keith Morgan

Andrew Passmore *Exeter Archaeology*

Andrew Pye *Archaeology Officer, Planning Services, Exeter City Council*

Edward Wirth *Archivist, US Dept. of Interior, Edison National Historic Site, USA*

Peter Weddell *Head of Exeter Archaeology*

Tyne & Wear Museums

BIBLIOGRAPHY

Gledhill, D. & Lamb, P.G   *Electricity in Taunton 1809-1948*
Somerset Industrial Archaeological Soc. 1986

Lamb, P.G.   *Electricity in Bristol 1863-1948*
The Historical Association (Bristol Branch)
1981

Delderfield, E.R.   *Cavalcade by Candlelight*
W.J. Delderfield & Sons, 1950

Garcke, Emile   *Manuals of Electrical Undertakings*
Various years

Allen, C.   *A History of Electricity Generation &
Supply in Exeter 1882 - 1905*
Privately published 1986

Sambourne, R.C.   *Exeter – A Century of Public transport*
Glasney Press, Falmouth 1976

The Electricity Council   *History of Electricity Supply in Great Britain*
Electricity Council Intelligence Section, 1971

B.E.D.A., Inc.   *Michael Faraday. His life and work*
British Electrical Development Assoc., Inc.

Kenneth R. Swan   *Sir Joseph Swan*
"Science in Britain" series
Longmans, Green & Co., 1946

**Exeter Archaeology &
Keystone Historic
Buildings Consultants**   *Archaeological, Historical and Conservation
Study of the Exeter Canal Basin*
Exeter Archaeology Report No 00.18, 2000

South Western Electricity Historical Society

Lynmouth Hydro Scheme 1890

The South Western Electricity Historical Society has been pleased to assist Dick Passmore in his researches and is interested in saving stories of the old electrical undertakings of the Southwest.

The Society has a major electricity archive in Bristol enabling us to answer electrical historical enquiries from all over the globe through our web site. *(see below)*

The subscription is £7.50 per annum (Year 2008). The Society's current membership stands at around 140 throughout the Southwest Peninsula and beyond.

For electrical history enquiries telephone Peter Lamb on:

01275 463160

To join the Society, please access our website and download an application form:

www.swehs.co.uk

INDEX

(Excluding Appendices)

NOTE: Neither "Exeter Power Station", nor "the power station", is included in the Index, as both are mentioned on the majority of pages. Likewise, Exeter City Council is not included if it is merely referred to as "the council" – for the same reason.

Items in italics denote a photograph.